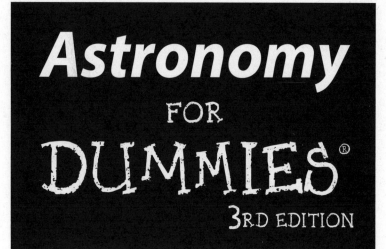

Astronomy

FOR

DUMMIES®

3RD EDITION

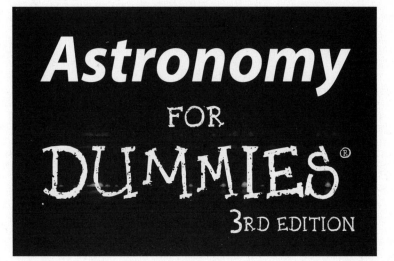

Astronomy FOR DUMMIES®
3RD EDITION

by Stephen P. Maran, PhD

WILEY

John Wiley & Sons, Inc.

Astronomy For Dummies®, 3rd Edition

Published by
John Wiley & Sons, Inc.
111 River St.
Hoboken, NJ 07030-5774
www.wiley.com

WILEY

About the Author

Stephen P. Maran, PhD, a 36-year veteran of the space program, has been honored with the NASA Medal for Exceptional Achievement. He received the Klumpke-Roberts Award of the Astronomical Society of the Pacific for "outstanding contributions to the public understanding and appreciation of astronomy," the George Van Biesbroeck Prize of the American Astronomical Society, for "long-term extraordinary or unselfish service to astronomy," and the Andrew W. Gemant Award of the American Institute of Physics for "significant contributions to the cultural, artistic, or humanistic dimension of physics."

In 2000, the International Astronomical Union named an asteroid (Minor Planet 9768) "Stephenmaran" for Dr. Maran.

Dr. Maran began practicing astronomy from rooftops in Brooklyn and at a deserted golf course in the far reaches of the Bronx. He graduated to conduct professional research with telescopes at the Kitt Peak National Observatory in Arizona, the National Radio Astronomy Observatory in West Virginia, the Palomar Observatory in California, and the Cerro Tololo Inter-American Observatory in Chile. He also conducted research with instruments in space, including the Hubble Space Telescope and the International Ultraviolet Explorer. He helped design and develop two instruments that flew in space aboard Hubble.

He's also taught astronomy at the University of California, Los Angeles, and the University of Maryland, College Park. As press officer of the American Astronomical Society for many years, he presided over media briefings that brought news of astronomical discoveries to people worldwide.

Dr. Maran observed total eclipses of the sun from the Gaspé Peninsula and elsewhere in Quebec; Baja California in Mexico; the Sahara desert in Libya; at sea off New Caledonia and Singapore in the eastern Pacific; and the shores of the U.S.

In the course of spreading the good word about astronomy, Maran has lectured on black holes in a bar in Tahiti and explained an eclipse of the Sun on NBC's show. He's also spoken on eclipse and comet cruises aboard Cunard's *Elizabeth 2*, Sitmar Line's *Fairwind*, and MSC Cruises' *Sinfonia*. He's addressed audiences ranging from inner-city Seattle school children and Atherton, California, Girl Scouts to the National Academy of Engineering in Washington, D.C. and subcommittees of both the U.S. House of Representatives and the U.N. Committee on the Peaceful Uses of Outer Space.

Dr. Maran is the author, coauthor, or editor of ten other books on astronomy and astrophysics. He has written many articles for national magazines and served as a consultant for the National Geographic Society, Time-Life Books, and other organizations.

Dr. Maran is a graduate of Stuyvesant High School in New York City (where he served a full season on the Math Team without incurring serious injury) and of Brooklyn College. He received both his MA and PhD in astronomy from the University of Michigan. Maran is married to Sally Scott Maran, a journalist. They have three children.

Dedication

To Sally, Michael, Enid, and Elissa with all my love.

Author's Acknowledgments

Thanks first to my family and friends who put up with me in the writing of this book. Thanks also to my agent, Skip Barker, who goaded and guided me in this project, and to Stacy Collins for her faith in the original project and Lindsay Lefevre for championing this substantially revised edition.

I'm grateful to Ron Cowen and Dr. Seth Shostak for their past contributions to this book; to Kathy Cox, Georgette Beatty, Josh Dials, and Jennifer Moore, who organized and edited it; and to their skilled colleagues on the editorial and production teams at Wiley Publishing who made the book better and brighter. A special thanks to Laurence Marschall for suggestions that improved the accuracy and thoroughness of the book.

Thanks as well to the organizations that provided the photographs in this book and to the producer of the star maps, Robert Miller.

I'm grateful for helpful advice from several experts for this new edition, particularly to Prof. Richard Mushotzky on black holes, Drs. Marc Kuchner and Rory Barnes on exoplanets, and Dr. Shostak on the search for extraterrestrial intelligence.

Some drawings in this book have been inspired by Dr. Dinah L. Moche and her excellent book, also published by Wiley. Dr. Moche deserves many thanks for her support of this book and for her dedication to making the science of astronomy accessible to anyone.

Publisher's Acknowledgments

We're proud of this book; please send us your comments at http://dummies.custhelp.com. For other comments, please contact our Customer Care Department within the U.S. at 877-762-2974, outside the U.S. at 317-572-3993, or fax 317-572-4002.

Some of the people who helped bring this book to market include the following:

Acquisitions, Editorial, and Vertical Websites

Project Editor: Jennifer Moore

Acquisitions Editor: Lindsay Lefevre

Copy Editor: Krista Hansing

Assistant Editor: David Lutton

Editorial Program Coordinator: Joe Niesen

Technical Editor: Laurence A. Marschall

Editorial Manager: Carmen Krikorian

Senior Editorial Manager: Jennifer Ehrlich

Editorial Assistants: Alexa Koschier and Rachelle Amick

Art Coordinator: Alicia B. South

Cover Photos:

Cartoons: Rich Tennant (www.the5thwave.com)

Composition Services

Project Coordinator: Sheree Montgomery

Layout and Graphics: Corrie Niehaus

Proofreader: Toni Settle

Indexer: Joan Griffitts

Publishing and Editorial for Consumer Dummies

    Kathleen Nebenhaus, Vice President and Executive Publisher

    David Palmer, Associate Publisher

    Kristin Ferguson-Wagstaffe, Product Development Director

Publishing for Technology Dummies

    Andy Cummings, Vice President and Publisher

Composition Services

    Debbie Stailey, Director of Composition Services

Contents at a Glance

Table of Contents

Introduction

· ·

*A*stronomy is the study of the sky, the science of cosmic objects and celestial happenings. It's nothing less than the investigation of the nature of the universe we live in. Astronomers carry out the business of astronomy by looking and (for radio astronomers) listening, using backyard telescopes, huge observatory instruments, and satellites orbiting Earth or positioned in space near Earth or another celestial body, such as the Moon or a planet. Scientists send up telescopes in sounding rockets and on unmanned balloons, some instruments travel far into the solar system aboard deep space probes, and some probes gather samples and return them to Earth.

Astronomy can be a professional or amateur activity. About 20,000 professional astronomers engage in space science worldwide, and an estimated 500,000 amateur astronomers live around the globe. Many of the amateurs belong to local or national astronomy clubs in their home countries.

Professional astronomers conduct research on the Sun and the solar system, the Milky Way galaxy, and the universe beyond. They teach in universities, design satellites in government labs, and operate planetariums. They also write books like this one (but maybe not as good). Most hold PhDs. Nowadays, many professional astronomers study abstruse physics of the cosmos or work with automated, remotely controlled telescopes, so they may not even know the constellations.

Amateur astronomers know the constellations. They share an exciting hobby. Some stargaze on their own; many others join astronomy clubs and organizations of every description. The clubs pass on know-how from old hands to new members, share telescopes and equipment, and hold meetings where members tell about their recent observations or hear lectures by visiting scientists.

Amateur astronomers also hold observing meetings where everyone brings a telescope (or looks through another observer's scope). The amateurs conduct these sessions at regular intervals (such as the first Saturday night of each month) or on special occasions (such as the return of a major meteor shower each August or the appearance of a bright comet like Hale-Bopp). And they save up for really big events, such as a total eclipse of the Sun, when thousands of amateurs and dozens of pros travel across Earth to position themselves in the path of totality and witness one of nature's greatest spectacles.

About This Book

This book explains all you need to know to launch into the great hobby of astronomy. It gives you a leg up on understanding the basic science of the universe as well. The latest space missions will make more sense to you: You'll understand why NASA and other organizations send space probes to planets like Saturn, why robot rovers land on Mars, and why scientists seek samples of the dust in the tail of a comet. You'll know why the Hubble Space Telescope peers out into space and how to check up on other space missions. And when astronomers show up in the newspaper or on television to report their latest discoveries — from space; from the big telescopes in Arizona, Hawaii, Chile, and California; or from radio telescopes in New Mexico, Puerto Rico, Australia, or other observatories around the world — you'll understand the background and appreciate the news. You'll even be able to explain it to your friends.

Read only the parts you want, in any order you want. I explain what you need as you go. Astronomy is fascinating and fun, so keep reading. Before you know it, you'll be pointing out Jupiter, spotting famous constellations and stars, and tracking the International Space Station as it whizzes by overhead. The neighbors may start calling you "stargazer." Police officers may ask you what you're doing in the park at night or why you're standing on the roof with binoculars. Tell 'em you're an astronomer. They probably haven't heard that one (I hope they believe you!).

Conventions Used in This Book

To help you navigate this book as you begin to navigate the skies, I use the following conventions:

- ✔ New or unfamiliar terms appear in *italics* and are accompanied by a concise definition.
- ✔ **Bold** indicates the action to take in numbered steps and highlights the key words in bulleted lists.
- ✔ Web addresses are set in `monofont` so you can easily spot them.

What You're Not to Read

Feel free to skip the sidebars that appear throughout the book; these shaded gray boxes contain interesting info that isn't essential to your understanding of astronomy. The same goes for any text I mark with the Technical Stuff icon.

Foolish Assumptions

You may be reading this book because you want to know what's up in the sky or what the scientists in the space program are doing. Perhaps you've heard that astronomy is a neat hobby, and you want to see whether the rumor is true. Perhaps you want to find out what equipment you need.

You're not a scientist. You just enjoy looking at the night sky and have fallen under its spell, wanting to see and understand the real beauty of the universe.

You want to observe the stars, but you also want to know what you're seeing. Maybe you even want to make a discovery of your own. You don't have to be an astronomer to spot a new comet, and you can even help listen for E.T. Whatever your goal, this book helps you achieve it.

Icons Used in This Book

Throughout this book, helpful icons highlight particularly useful information — even if they just tell you to not sweat the tough stuff. Here's what each symbol means.

Observation is the key to astronomy, and these tips help make you an observational pro. I help you scope out techniques and opportunities to fine-tune your viewing technique.

This nerdy guy appears beside discussions that you can skip if you just want to know the basics and start watching the skies. The scientific background can be good to know, but many people happily enjoy their stargazing without knowing about the physics of supernovas, the mathematics of galaxy chasing, and the ins and outs of dark energy.

This target puts you right on track to make use of some inside information as you start skywatching or make progress in the hobby.

How much trouble can you get into watching the stars? Not much, if you're careful. But some things you can't be too careful about. This bomb alerts you to pay attention so you don't get burned.

Where to Go from Here

You can start anywhere you want. Worried about the fate of the universe? Start off with the Big Bang (see Chapter 16 if you're really interested).

Or you may want to begin with what's in store for you as you pursue your passion for the stars.

Wherever you start, I hope you continue your cosmic exploration and experience the joy, excitement, enlightenment, and enchantment that people have always found in the skies.

Part I
Stalking the Cosmos

GEORGE NEVER RULED OUT THE POSSIBILITY OF EXTRATERRESTRIAL LIFE

ANDROMEDA'S GREATEST DAD

In this part . . .

Objects and events in the sky have always fascinated us humans. Throughout history, our interest in astronomy has been both practical and inspired. People navigated by the stars and planted crops according to the Moon's phases (and still do today). They built sites where observations may have accompanied rituals (such as Stonehenge) and kept time by the motions of the Sun and stars. And folks still wonder about the nature of objects in the heavens.

You can join this grand human tradition. In this part, I introduce the science of astronomy and offer techniques and advice for observing planets, comets, meteors, and other sights in the night sky.

Chapter 1

Seeing the Light: The Art and Science of Astronomy

Step outside on a clear night and look at the sky. If you're a city dweller or live in a cramped suburb, you see dozens, maybe hundreds, of twinkling stars. Depending on the time of the month, you may also see a full Moon and up to five of the eight planets that revolve around the Sun.

A shooting star or "meteor" may appear overhead. What you actually see is the flash of light from a tiny piece of comet dust streaking through the upper atmosphere.

Another pinpoint of light moves slowly and steadily across the sky. Is it a space satellite, such as the Hubble Space Telescope, or just a high-altitude airliner? If you have a pair of binoculars, you may be able to see the difference. Most airliners have running lights, and their shapes may be perceptible.

If you live in the country — on the seashore away from resorts and developments, on the plains, or in the mountains far from any floodlit ski slope — you can see thousands of stars. The Milky Way appears as a beautiful pearly swath across the heavens. What you're seeing is the cumulative glow from millions of faint stars, individually indistinguishable with the naked eye. At a great observation place, such as Cerro Tololo in the Chilean Andes, you can see even more stars. They hang like brilliant lamps in a coal black sky, often not even twinkling, like in van Gogh's *Starry Night* painting.

When you look at the sky, you practice astronomy — you observe the universe that surrounds you and try to make sense of what you see. For thousands of years, everything people knew about the heavens they deduced by simply observing the sky. Almost everything that astronomy deals with

✔ Is seen from a distance

✔ Is discovered by studying the light that comes to you from objects in space

✔ Moves through space under the influence of gravity

This chapter introduces you to these concepts (and more).

Astronomy: The Science of Observation

Astronomy is the study of the sky, the science of cosmic objects and celestial happenings, and the investigation of the nature of the universe you live in. Professional astronomers carry out the business of astronomy by observing with telescopes that capture visible light from the stars or by tuning in to radio waves that come from space. They use backyard telescopes, huge observatory instruments, and satellites that orbit Earth collecting forms of light (such as ultraviolet radiation) that the atmosphere blocks from reaching the ground. They send up telescopes in sounding rockets (equipped with instruments for making high-altitude scientific observations) and on unmanned balloons. And they send some instruments into the solar system aboard deep-space probes.

Professional astronomers study the Sun and the solar system, the Milky Way, and the universe beyond. They teach in universities, design satellites in government labs, and operate planetariums. They also write books (like me, your loyal *For Dummies* hero). Most have completed years of schooling to hold PhDs. Many of them study complex physics or work with automated, robotic telescopes that reach far beyond the night sky recognizable to our eyes. They may not have ever studied the *constellations* (groups of stars, such as Ursa Major, the Great Bear, named by ancient stargazers) that amateur or hobbyist astronomers first explore.

You may already be familiar with the Big Dipper, an asterism in Ursa Major. An *asterism* is a named star pattern that's not identical to one of the 88 recognized constellations. An asterism may be wholly within a single constellation or may include stars from more than one constellation. For example, the four corners of the Great Square of Pegasus, a large asterism, are marked by three stars of the Pegasus constellation and a fourth from Andromeda. Figure 1-1 shows the Big Dipper in the night sky. (In the United Kingdom, some people call the Big Dipper *the Plough*.)

Figure 1-1:
The Big
Dipper,
found in
Ursa Major,
is an
asterism.

In addition to the roughly 20,000 professional astronomers worldwide, several hundred thousand amateur astronomers enjoy watching the skies. Amateur astronomers usually know the constellations and use them as guideposts when exploring the sky by eye, with binoculars, and with telescopes. Many amateurs also make useful scientific contributions. They monitor the changing brightness of variable stars; discover asteroids, comets, and exploding stars; and crisscross Earth to catch the shadows cast as asteroids pass in front of bright stars (thereby helping astronomers map the asteroids' shapes). They even join in professional research efforts with their home computers and smartphones through Citizen Science projects, which I describe in Chapter 2 and elsewhere throughout the book.

In the rest of Part I, I provide you with information on how to observe the skies effectively and enjoyably.

What You See: The Language of Light

Light brings us information about the planets, moons, and comets in our solar system; the stars, star clusters, and nebulae in our galaxy; and the objects beyond.

In ancient times, folks didn't think about the physics and chemistry of the stars; they absorbed and passed down folk tales and myths: the Great Bear, the Demon star, the Man in the Moon, the dragon eating the Sun during a solar eclipse, and more. The tales varied from culture to culture. But many people did discover the patterns of the stars. In Polynesia, skilled navigators rowed across hundreds of miles of open ocean with no landmarks in view and no compass. They sailed by the stars, the Sun, and their knowledge of prevailing winds and currents.

Gazing at the light from a star, the ancients noted its brightness, position in the sky, and color. This information helps people distinguish one sky object from another, and the ancients (and now people today) got to know them like old friends. Some basics of recognizing and describing what you see in the sky are

- ✔ Distinguishing stars from planets
- ✔ Identifying constellations, individual stars, and other sky objects by name
- ✔ Observing brightness (given as magnitudes)
- ✔ Understanding the concept of a light-year
- ✔ Charting sky position (measured in special units called *RA* and *Dec*)

They wondered as they wandered: Understanding planets versus stars

The term *planet* comes from the ancient Greek word *planetes,* meaning "wanderer." The Greeks (and other ancient people) noticed that five spots of light moved across the pattern of stars in the sky. Some moved steadily ahead; others occasionally looped back on their own paths. Nobody knew why. And these spots of light didn't twinkle like the stars did — no one understood that difference, either. Every culture had a name for those five spots of light — what we now call planets. Their English names are Mercury, Venus, Mars, Jupiter, and Saturn. These celestial bodies aren't wandering through the stars; they orbit around the Sun, our solar system's central star.

Today astronomers know that planets can be smaller or bigger than Earth, but they all are much smaller than the Sun. The planets in our solar system are so close to Earth that they have perceptible disks — at least, when viewed through a telescope — so we can see their shapes and sizes. The stars are so far away from Earth that even if you view them through a powerful telescope, they show up only as points of light. (For more about the planets in the solar system, flip to Part II.)

If you see a Great Bear, start worrying: Naming stars and constellations

I used to tell planetarium audiences who craned their necks to look at stars projected above them, "If you can't see a Great Bear up there, don't worry. Maybe those who *do* see a Great Bear should worry."

Ancient astronomers divided the sky into imaginary figures, such as Ursa Major (Latin for Great Bear); Cygnus, the Swan; Andromeda, the Chained Lady; and Perseus, the Hero. The ancients identified each figure with a pattern of stars. The truth is, to most people, Andromeda doesn't look much like a chained lady at all — or anything else, for that matter (see Figure 1-2).

Today astronomers have divided the sky into 88 constellations, which contain all the stars you can see. The International Astronomical Union, which governs the science, set boundaries for the constellations so that astronomers can agree on which star is in which constellation. Previously, sky maps drawn by different astronomers often disagreed. Now when you read that the Tarantula Nebula is in Dorado (see Chapter 12), you know that, to see this nebula, you must seek it in the Southern Hemisphere constellation Dorado, the Goldfish.

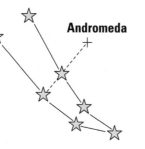

Andromeda

Figure 1-2: Andromeda is also known as the Chained Lady.

The largest constellation is Hydra, the Water Snake. The smallest is Crux, the Cross, which most people call the Southern Cross. You can see a Northern Cross, too, but you can't find it in a list of constellations; it's an asterism (a term defined previously) within Cygnus, the Swan. Although astronomers generally agree on the names of the constellations, they don't have a consensus on what each name means. For example, some astronomers call Dorado the Swordfish, but I'd like to spike that name. One constellation, Serpens, the Serpent, is broken into two sections that aren't connected. The two sections, located on either side of Ophiuchus, the Serpent Bearer, are Serpens Caput (the Serpent's Head) and Serpens Cauda (the Serpent's Tail).

The individual stars in a constellation often have no relation to each other except for their proximity in the sky as visible from Earth. In space, the stars that make up a constellation may be completely unrelated to one another, with some located relatively near Earth and others located at much greater distances in space. But they make a simple pattern for observers on Earth to enjoy.

As a rule, the brighter stars in a constellation were assigned a Greek letter, either by the ancient Greeks or by astronomers of later civilizations. In each constellation, the brightest star was labeled alpha, the first letter of the Greek alphabet.

The next brightest star was beta, the second Greek letter, and so on down to omega, the final letter of the 24-character Greek alphabet. (The astronomers used only lowercase Greek letters, so you see them written as α, β, . . . ω.)

So Sirius, the brightest star in the night sky — in Canis Major, the Great Dog — is called Alpha Canis Majoris. (Astronomers add a suffix here or there to put star names in the Latin genitive case — scientists have always liked Latin.) Table 1-1 shows a list of the Greek alphabet, in order, with the names of the letters and their corresponding symbols.

Table 1-1	The Greek Alphabet
Letter	*Name*
α	Alpha
β	Beta
γ	Gamma
δ	Delta
ε	Epsilon
ζ	Zeta
η	Eta
θ	Theta
ι	Iota
κ	Kappa
λ	Lambda
μ	Mu
ν	Nu
ξ	Xi
ο	Omicron
π	Pi
ρ	Rho
σ	Sigma
τ	Tau
υ	Upsilon
φ	Phi
χ	Chi
ψ	Psi
ω	Omega

When you look at a star atlas, you discover that the individual stars in a constellation aren't marked α Canis Majoris, β Canis Majoris, and so on. Usually, the creator of the atlas marks the area of the whole constellation as Canis Major and labels the individual stars α, β, and so on. When you read about a star in a list of objects to observe, say, in an astronomy magazine (see Chapter 2), you probably won't see it listed in the style of Alpha Canis Majoris or even α Canis Majoris. Instead, to save space, the magazine prints it as α CMa; *CMa* is the three-letter abbreviation for Canis Majoris (and also the abbreviation for Canis Major). I give the abbreviation for each of the constellations in Table 1-2.

Astronomers didn't coin special names such as Sirius for every star in Canis Major, so they named them with Greek letters or other symbols. In fact, some constellations don't have a single named star. (Don't fall for those advertisements that offer to name a star for a fee. The International Astronomical Union doesn't recognize purchased star names.) In other constellations, astronomers assigned Greek letters, but they could see more stars than the 24 Greek letters. Therefore, astronomers gave some stars Arabic numbers or letters from the Roman alphabet, or numbers in professional catalogues. So you see star names such as 61 Cygni, b Vulpeculae, HR 1516, and more. You may even run across the star names RU Lupi and YY Sex. (I'm not making this up.) But as with any other star, you can recognize them by their positions in the sky (as tabulated in star lists), their brightness, their color, or other properties, if not their names.

When you look at the constellations today, you see many exceptions to the rule that the Greek-letter star names correspond to the respective brightness of the stars in a constellation. The exceptions exist because

- ✔ The letter names were based on inaccurate naked-eye observations of brightness.

- ✔ Over the years, star atlas authors changed constellation boundaries, moving some stars from one constellation into another that included previously named stars.

- ✔ Some astronomers mapped out small and Southern Hemisphere constellations long after the Greek period, and they didn't always follow the lettering practice.

- ✔ The brightness of some stars has changed over the centuries since the ancient Greeks charted them.

A good (or bad) example is the constellation Vulpecula, the Fox, in which only one of the stars (alpha) has a Greek letter.

Because alpha isn't always the brightest star in a constellation, astronomers needed another term to describe that exalted status, and *lucida* is the word (from the Latin word *lucidus,* meaning "bright" or "shining"). The lucida of

Canis Major is Sirius, the alpha star, but the lucida of Orion, the Hunter, is Rigel, which is Beta Orionis. The lucida of Leo Minor, the Little Lion (a particularly inconspicuous constellation), is 46 Leo Minoris.

Table 1-2 lists the 88 constellations, the brightest star in each, and the magnitude of that star. *Magnitude* is a measure of a star's brightness. (I talk about magnitudes later in this chapter in the section "The smaller, the brighter: Getting to the root of magnitudes.") When the lucida of a constellation is the alpha star and has a name, I list only the name. For example, in Auriga, the Charioteer, the brightest star, Alpha Aurigae, is Capella. But when the lucida isn't an alpha, I give its Greek letter or other designation in parentheses. For example, the lucida of Cancer, the Crab, is Al Tarf, which is Beta Cancri.

Table 1-2		The Constellations and Their Brightest Stars		
Name	*Abbreviation*	*Meaning*	*Star*	*Magnitude*
Andromeda	And	Chained Lady	Alpheratz	2.1
Antlia	Ant	Air Pump	Alpha Antliae	4.3
Apus	Aps	Bird of Paradise	Alpha Apodis	3.8
Aquarius	Aqr	Water Bearer	Sadalmelik	3.0
Aquila	Aql	Eagle	Altair	0.8
Ara	Ara	Altar	Beta Arae	2.9
Aries	Ari	Ram	Hamal	2.0
Auriga	Aur	Charioteer	Capella	0.1
Bootes	Boo	Herdsman	Arcturus	−0.04
Caelum	Cae	Chisel	Alpha Caeli	4.5
Camelopardalis	Cam	Giraffe	Beta Camelopardalis	4.0
Cancer	Cnc	Crab	Al Tarf (Beta Cancri)	3.5
Canes Venatici	CVn	Hunting Dogs	Cor Caroli	2.8
Canis Major	CMa	Great Dog	Sirius	−1.5
Canis Minor	CMi	Little Dog	Procyon	0.4
Capricornus	Cap	Goat	Deneb Algedi (Delta Capricorni)	2.9
Carina	Car	Ship's Keel	Canopus	−0.7
Cassiopeia	Cas	Queen	Schedar	2.2
Centaurus	Cen	Centaur	Rigil Kentaurus	−0.3

Name	Abbreviation	Meaning	Star	Magnitude
Cepheus	Cep	King	Alderamin	2.4
Cetus	Cet	Whale	Deneb Kaitos (Beta Ceti)	2.0
Chamaeleon	Cha	Chameleon	Alpha Chamaeleontis	4.1
Circinus	Cir	Compasses	Alpha Circini	3.2
Columba	Col	Dove	Phakt	2.6
Coma Berenices	Com	Berenice's Hair	Beta Comae Berenices	4.3
Corona Australis	CrA	Southern Crown	Alpha Coronae Australis	4.1
Corona Borealis	CrB	Northern Crown	Alphekka	2.2
Corvus	Crv	Crow	Gienah (Gamma Corvi)	2.6
Crater	Crt	Cup	Delta Crateris	3.6
Crux	Cru	Cross	Acrux	0.7
Cygnus	Cyg	Swan	Deneb	1.3
Delphinus	Del	Dolphin	Rotanev (Beta Delphini)	3.6
Dorado	Dor	Goldfish	Alpha Doradus	3.3
Draco	Dra	Dragon	Thuban	3.7
Equuleus	Equ	Little Horse	Kitalpha	3.9
Eridanus	Eri	River	Achernar	0.5
Fornax	For	Furnace	Alpha Fornacis	3.9
Gemini	Gem	Twins	Pollux (Beta Geminorum)	1.1
Grus	Gru	Crane	Alnair	1.7
Hercules	Her	Hercules	Ras Algethi	2.6
Horologium	Hor	Clock	Alpha Horologii	3.9
Hydra	Hya	Water Snake	Alphard	2.0
Hydrus	Hyi	Little Water Snake	Beta Hydri	2.8
Indus	Ind	Indian	Alpha Indi	3.1
Lacerta	Lac	Lizard	Alpha Lacertae	3.8
Leo	Leo	Lion	Regulus	1.4

(continued)

Table 1-2 *(continued)*

Name	Abbreviation	Meaning	Star	Magnitude
Leo Minor	LMi	Little Lion	Praecipua (46 Leo Minoris)	3.8
Lepus	Lep	Hare	Arneb	2.6
Libra	Lib	Scales	Zubeneschemali (Beta Librae)	2.6
Lupus	Lup	Wolf	Alpha Lupus	2.3
Lynx	Lyn	Lynx	Alpha Lyncis	3.1
Lyra	Lyr	Lyre	Vega	0.0
Mensa	Men	Table	Alpha Mensae	5.1
Microscopium	Mic	Microscope	Gamma Microscopii	4.7
Monoceros	Mon	Unicorn	Beta Monocerotis	3.7
Musca	Mus	Fly	Alpha Muscae	2.7
Norma	Nor	Level and Square	Gamma Normae	4.0
Octans	Oct	Octant	Nu Octantis	3.8
Ophiuchus	Oph	Serpent Bearer	Rasalhague	2.1
Orion	Ori	Hunter	Rigel (Beta Orionis)	0.1
Pavo	Pav	Peacock	Alpha Pavonis	1.9
Pegasus	Peg	Winged Horse	Enif (Epsilon Pegasi)	2.4
Perseus	Per	Hero	Mirphak	1.8
Phoenix	Phe	Phoenix	Ankaa	2.4
Pictor	Pic	Easel	Alpha Pictoris	3.2
Pisces	Psc	Fish	Eta Piscium	3.6
Pisces Austrinus	PsA	Southern Fish	Fomalhaut	1.2
Puppis	Pup	Ship's Stern	Zeta Puppis	2.3
Pyxis	Pyx	Compass	Alpha Pyxidus	3.7
Reticulum	Ret	Net	Alpha Reticuli	3.4
Sagitta	Sge	Arrow	Gamma Sagittae	3.5

Name	Abbreviation	Meaning	Star	Magnitude
Sagittarius	Sgr	Archer	Kaus Australis (Epsilon Sagittarii)	1.9
Scorpius	Sco	Scorpion	Antares	1.0
Sculptor	Scl	Sculptor	Alpha Sculptoris	4.3
Scutum	Sct	Shield	Alpha Scuti	3.9
Serpens	Ser	Serpent	Unukalhai	2.7
Sextans	Sex	Sextant	Alpha Sextantis	4.5
Taurus	Tau	Bull	Aldebaran	0.9
Telescopium	Tel	Telescope	Alpha Telescopium	3.5
Triangulum	Tri	Triangle	Beta Trianguli	3.0
Triangulum Australe	TrA	Southern Triangle	Alpha Trianguli Australis	1.9
Tucana	Tuc	Toucan	Alpha Tucanae	2.9
Ursa Major	UMa	Great Bear	Alioth (Epsilon Ursae Majoris)	1.8
Ursa Minor	UMi	Little Bear	Polaris	2.0
Vela	Vel	Sails	Suhail al Muhlif (Gamma Velorum)	1.8
Virgo	Vir	Virgin	Spica	1.0
Volans	Vol	Flying Fish	Gamma Volantis	3.6
Vulpecula	Vul	Fox	Anser	4.4

Identifying stars would be much easier if they had little name tags that you could see through your telescope, but at least they don't have unlisted numbers like Hollywood celebrities. If you have a smartphone, you can download an app to identify the stars for you. Just download a sky map or planetarium app (such as Sky Safari, Star Walk, or Google Sky Map) and face the phone toward the sky. The app generates a map of the constellations in the general direction your phone is facing. With some apps, when you touch the image of a star, its name appears. (I describe more astronomy apps in Chapter 2; for the full scoop on stars, check out Chapter 11.)

What do I spy? Spotting the Messier Catalog and other sky objects

Naming stars was easy enough for astronomers. But what about all those other objects in the sky — galaxies, nebulae, star clusters, and the like (which I cover in Part III)? Charles Messier (1730–1817), a French astronomer, created a numbered list of about 100 fuzzy sky objects. His list is known as the *Messier Catalog,* and now when you hear the Andromeda Galaxy called by its scientific name, M31, you know that it stands for number 31 in the Catalog. Today 110 objects make up the standard Messier Catalog.

You can find pictures and a complete list of the Messier objects at The Messier Catalog website of Students for the Exploration and Development of Space at www.seds.org/messier. And you can find out how to earn a certificate for viewing Messier objects from the Astronomical League Messier Club website at www.astroleague.org/al/obsclubs/messier/mess.html.

Experienced amateur astronomers often engage in Messier marathons, in which each person tries to observe every object in the *Messier Catalog* during a single long night. But in a marathon, you don't have time to enjoy an individual nebula, star cluster, or galaxy. My advice is to take it slow and savor their individual visual delights. A wonderful book on the Messier objects, which includes hints on how to observe each object, is Stephen J. O'Meara's *Deep-Sky Companions: The Messier Objects* (Cambridge University Press, 2000).

Since Messier's time, astronomers have confirmed the existence of thousands of other *deep sky objects,* the term amateurs use for star clusters, nebulae, and galaxies to distinguish them from stars and planets. Because Messier didn't list them, astronomers refer to these objects by their numbers as given in other catalogues. You can find many of these objects listed in viewing guides and sky maps by their NGC *(New General Catalogue)* and IC *(Index Catalogue)* numbers. For example, the bright double cluster in Perseus, the Hero, consists of NGC 869 and NGC 884.

When you've become familiar with some of the Messier objects, consider tackling the Caldwell objects with your telescope and observing skills. Consult the Caldwell Program Object List, posted online by the Astronomical League at www.astroleague.org/al/obsclubs/caldwell/cldwlist.html. It provides basic information on 109 star clusters, nebulae, and galaxies selected for your viewing pleasure by British astronomer Sir Patrick Moore. When you observe this list, it's also helpful to consult *The Caldwell Objects and How to Observe Them,* by Martin Mobberley (Springer, 2009).

The smaller, the brighter: Getting to the root of magnitudes

A star map, constellation drawing, or list of stars always indicates each star's magnitude. The *magnitudes* represent the brightness of the stars. One of the ancient Greeks, Hipparchos (also spelled Hipparchus, but he wrote it in Greek), divided all the stars he could see into six classes. He called the brightest stars magnitude 1 or *1st magnitude,* the next brightest bunch the *2nd magnitude* stars, and on down to the dimmest ones, which were *6th magnitude.*

By the numbers: The mathematics of brightness

The 1st magnitude stars are about 100 times brighter than the 6th magnitude stars. In particular, the 1st magnitude stars are about 2.512 times brighter than the 2nd magnitude stars, which are about 2.512 times brighter than the 3rd magnitude stars, and so on. (At the 6th magnitude, you get up into some big numbers: 1st magnitude stars are about 100 times brighter.) You mathematicians out there recognize this as a *geometric progression.* Each magnitude is the 5th root of 100 (meaning that when you multiply a number by itself four times — for example, $2.512 \times 2.512 \times 2.512 \times 2.512 \times 2.512$ — the result is 100). If you doubt my word and do this calculation on your own, you get a slightly different answer because I left off some decimal places.

Thus, you can calculate how faint a star is — compared to some other star — from its magnitude. If two stars are 5 magnitudes apart (such as the 1st magnitude star and the 6th magnitude star), they differ by a factor of 2.512^5 (2.512 to the fifth power), and a good pocket calculator shows you that one star is 100 times brighter. If two stars are 6 magnitudes apart, one is about

250 times brighter than the other. And if you want to compare, say, a 1st magnitude star with an 11th magnitude star, you compute a 2.512^{10} difference in brightness, meaning a factor of 100^2, or 10,000.

The faintest object visible with the Hubble Space Telescope is about 25 magnitudes fainter than the faintest star you can see with the naked eye (assuming normal vision and viewing skills — some experts and a certain number of liars and braggarts say that they can see 7th magnitude stars). Speaking of dim stars, 25 magnitudes are five times 5 magnitudes, which corresponds to a brightness difference of a factor of 100^5. So the Hubble can see $100 \times 100 \times 100 \times 100 \times 100$, or 10 billions times fainter than the human eye. Astronomers expect nothing less from a billion-dollar telescope. At least it didn't cost $10 billion.

You can get a good telescope for well under $1,000, and you can view the billion-dollar Hubble's best photos on the Internet for free at www.hubblesite.org.

Notice that, contrary to most common measurement scales and units, the brighter the star, the smaller the magnitude. The Greeks weren't perfect, however; even Hipparchos had an Achilles' heel: He didn't leave room in his system for the very brightest stars, when accurately measured.

So today we recognize a few stars with a zero magnitude or a negative magnitude. Sirius, for example, is magnitude –1.5. And the brightest planet, Venus, is sometimes magnitude –4 (the exact value differs, depending on the distance Venus is from Earth at the time and its direction with respect to the Sun).

Another omission: Hipparchos didn't have a magnitude class for stars that were too dim to be seen with the naked eye. This didn't seem like an oversight at the time because nobody knew about these stars before the invention of the telescope. But today astronomers know that billions of stars exist beyond our naked-eye view. Their magnitudes are larger numbers: 7 or 8 for stars easily seen through binoculars, and 10 or 11 for stars easily seen through a good, small telescope. The magnitudes reach as high (and as dim) as 21 for the faintest stars in the Palomar Observatory Sky Survey and about 31 for the faintest objects imaged with the Hubble Space Telescope.

Looking back on light-years

The distances to the stars and other objects beyond the planets of our solar system are measured in *light-years*. As a measurement of actual length, a light-year is about 5.9 trillion miles long.

People confuse a light-year with a length of time because the term contains the word *year*. But a light-year is really a distance measurement — the length that light travels, zipping through space at 186,000 miles per second, over the course of a year.

When you view an object in space, you see it as it appeared when the light left the object. Consider these examples:

- ✔ When astronomers spot an explosion on the Sun, we don't see it in real time; the light from the explosion takes about 8 minutes to get to Earth.

- ✔ The nearest star beyond the Sun, Proxima Centauri, is about 4 light-years away. Astronomers can't see Proxima as it is now — only as it was four years ago.

- ✔ Look up at the Andromeda Galaxy, the most distant object that you can readily see with the unaided eye, on a clear, dark night in the fall. The light your eye receives left that galaxy about 2.6 million years ago. If there was a big change in Andromeda tomorrow, we wouldn't know that it happened for more than 2 million years. (See Chapter 12 for hints on viewing the Andromeda Galaxy and other prominent galaxies.)

TECHNICAL STUFF

Hey, you! No, no, I mean AU

Earth is about 93 million miles from the Sun, or 1 *astronomical unit* (AU). The distances between objects in the solar system are usually given in AU. Its plural is also AU. (Don't confuse AU with "Hey, you!")

In public announcements, press releases, and popular books, astronomers state how far the stars and galaxies that they study are "from Earth." But among themselves and in technical journals, they always give the distances from the Sun, the center of our solar system. This discrepancy rarely matters because astronomers can't measure the distances of the stars precisely enough for 1 AU more or less to make a difference, but they do it this way for consistency.

Here's the bottom line:

- When you look out into space, you're looking back in time.
- Astronomers don't have a way to know exactly what an object out in space looks like right now.

When you look at some big, bright stars in a faraway galaxy, you must entertain the possibility that those particular stars don't even exist anymore. Some massive stars live for only 10 million or 20 million years. If you see them in a galaxy that is 50 million light-years away, you're looking at lame duck stars. They aren't shining in that galaxy anymore; they're dead.

If astronomers send a flash of light toward one of the most distant galaxies found with Hubble and other major telescopes, the light would take billions of years to arrive. Astronomers, however, calculate that the Sun will swell up and destroy all life on Earth a mere 5 billion or 6 billion years from now, so the light would be a futile advertisement of our civilization's existence, a flash in the celestial pan.

Keep on moving: Figuring the positions of the stars

Astronomers used to call stars "fixed stars," to distinguish them from the wandering planets. But in fact, stars are in constant motion as well, both real and apparent. The whole sky rotates overhead because Earth is turning. The stars rise and set, like the Sun and the Moon, but they stay in formation. The stars that make up the Great Bear don't swing over to the Little Dog or Aquarius, the Water Bearer. Different constellations rise at different times and on different dates, as seen from different places around the globe.

Actually, the stars in Ursa Major (and every other constellation) do move with respect to one another — and at breathtaking speeds, measured in hundreds of miles per second. But those stars are so far away that scientists need precise measurements over considerable intervals of time to detect their motions across the sky. So 20,000 years from now, the stars in Ursa Major will form a different pattern in the sky. (Maybe they will even look like a Great Bear.)

In the meantime, astronomers have measured the positions of millions of stars, and many of them are tabulated in catalogs and marked on star maps. The positions are listed in a system called right ascension and declination — known to all astronomers, amateur and pro, as *RA* and *Dec:*

- The RA is the position of a star measured in the east–west direction on the sky (like longitude, the position of a place on Earth measured east or west of the prime meridian at Greenwich, England).

- The Dec is the position of the star measured in the north–south direction, like the latitude of a city, which is measured north or south of the equator.

Digging deeper into RA and Dec

A star at RA 2h00m00s is 2 hours east of a star at RA 0h00m00s, regardless of their declinations. RA increases from west to east, starting from RA 0h00m00s, which corresponds to a line in the sky (actually half a circle, centered on the center of Earth) from the North Celestial Pole to the South Celestial Pole. The first star may be at Dec 30° North, and the second star may be at Dec 15° 25'12" South, but they're still 2 hours apart in the east–west direction (and 45° 25'12" apart in the north–south direction). The North and South Celestial Poles are the points in the sky — due north and due south — around which the whole sky seems to turn, with the stars all rising and setting.

Note the following details about the units of RA and Dec:

- An hour of RA equals an arc of 15 degrees on the equator in the sky. Twenty-four hours of RA span the sky, and 24 × 15 = 360 degrees, or a complete circle around the sky. A minute of RA, called a *minute of time,* is a measure of angle on the sky that makes up 1/60 of an hour of RA. So you take 15° ÷ 60, or 1/4°. A second of RA, or a *second of time,* is 60 times smaller than a minute of time.

- Dec is measured in degrees, like the degrees in a circle, and in minutes and seconds of arc. A whole degree is about twice the apparent or angular size of the full Moon. Each degree is divided into 60 minutes of arc. The Sun and the full Moon are both about 32 minutes of arc (32') wide, as seen on the sky, although, in reality, the Sun is much larger than the Moon. Each minute of arc is divided into 60 seconds of arc (60"). When you look through a backyard telescope at high magnification, turbulence in the air blurs the image of a star. Under good conditions (low turbulence), the image should measure about 1" or 2" across. That's 1 or 2 arc seconds, not 1 or 2 inches.

Astronomers usually list RA in units of hours, minutes, and seconds, like time. We list Dec in degrees, minutes, and seconds of arc. Ninety degrees make up a right angle, 60 minutes of arc make up a degree, and 60 seconds of arc equal a minute of arc. A minute or second of arc is also often called an "arc minute" or an "arc second," respectively.

A few simple rules may help you remember how RA and Dec work and how to read a star map (see Figure 1-3):

- The North Celestial Pole (NCP) is the place to which the axis of Earth points in the north direction. If you stand at the geographic North Pole, the NCP is right overhead. (If you stand there, say "Hi" to Santa for me, but beware: You may be on thin ice, since there's no land at the geographic North Pole.)

- The South Celestial Pole (SCP) is the place to which the axis of Earth points in the south direction. If you stand at the geographic South Pole, the SCP is right overhead. I hope you dressed warmly: You're in Antarctica!

- The imaginary lines of equal RA run through the NCP and SCP as semicircles centered on the center of Earth. They may be imaginary, but they appear marked on most sky maps to help people find the stars at particular RAs.

- The imaginary lines of equal Dec, like the line in the sky that marks Dec of 30° North, pass overhead at the corresponding geographic latitudes. So if you stand in New York City, latitude 41° North, the point overhead is always at Dec 41° North, although its RA changes constantly as Earth turns. These imaginary lines appear on star maps, too, as *declination circles*.

Suppose you want to find the NCP as visible from your backyard. Face due north and look at an altitude of *x* degrees, where *x* is your geographic latitude. I'm assuming that you live in North America, Europe, or somewhere in the Northern Hemisphere. If you live in the Southern Hemisphere, you can't see the NCP. You can, however, look for the SCP. Look for the spot due south whose altitude in the sky, measured in degrees above the horizon, is equal to your geographic latitude.

In almost every astronomy book, the symbol " means seconds of arc, not inches. But at every university, a student in Astronomy 101 always writes on an exam, "The image of the star was about 1 inch in diameter." Understanding beats memorizing every day, but not everyone understands.

Here's the good news: If you just want to spot the constellations and the planets, you don't have to know how to use RA and Dec. Just consult a star map drawn for the current week or month (you can find these on the website of *Sky & Telescope* or one of the other magazines that I mention in Chapter 2, in the magazines themselves, or using a desktop planetarium program for your home computer or a planetarium app for your smartphone or tablet; I recommend programs, websites, and apps in Chapter 2 as well). But if you want to understand how star catalogs and maps work and how to zero in on faint galaxies with your telescope, understanding the system helps.

And if you purchase one of those snazzy, new, surprisingly affordable telescopes with computer control (see Chapter 3), you can punch in the RA and Dec of a recently discovered comet, and the scope points right at it. (A little table called an *ephemeris* comes with every announcement of a new comet. It gives the predicted RA and Dec of the comet on successive nights as it sweeps across the sky.)

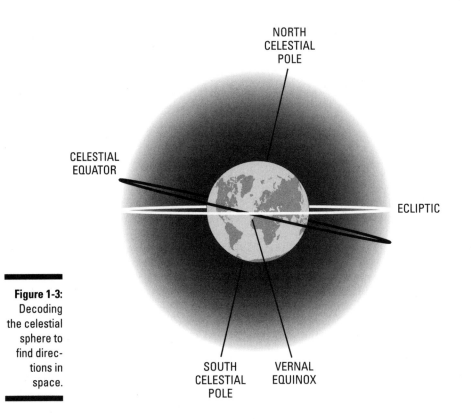

Figure 1-3:
Decoding
the celestial
sphere to
find direc-
tions in
space.

Gravity: A Force to Be Reckoned With

Ever since the work of Sir Isaac Newton, the English scientist (1642–1727), everything in astronomy has revolved around gravity. Newton explained gravity as a force between any two objects. The force depends on mass and separation. The more massive the object, the more powerful its pull. The greater the distance, the weaker the gravitational attraction. Newton sure was a smart cookie!

Albert Einstein developed an improved theory of gravity, which passes experimental tests that Newton's theory flunks. Newton's theory was good enough for commonly experienced gravity, like the force that made the apple fall on his head (if it really hit him). But in other respects, Newton's theory was hit or miss. Einstein's theory is better because it predicts everything that Newton got right, but also predicts effects that happen close to massive objects, where gravity is very strong. Einstein didn't think of gravity as a force; he considered it the bending of space and time by the very presence of a massive object, such as a star. I get all bent out of shape just thinking about it.

Newton's concept of gravity explains the following:

- ✔ Why the Moon orbits Earth, why Earth orbits the Sun, why the Sun orbits the center of the Milky Way, and why many other objects orbit one object or another out there in space

- ✔ Why a star or a planet is round

- ✔ Why gas and dust in space may clump together to form new stars

Einstein's theory of gravity, the General Theory of Relativity, explains the following:

- ✔ Why stars visible near the Sun during a total eclipse seem slightly out of position

- ✔ Why black holes exist

- ✔ Why gravitational lensing is found when we observe deep space

- ✔ Why, as Earth turns, it drags warped space and time around with it, an effect that scientists have verified with the help of satellites orbiting Earth

You find out about black holes in Chapters 11 and 13, and you can read up on gravitational lensing in Chapters 11, 14, and 15 without mastering the General Theory of Relativity.

You'll get smarter if you read every chapter in this book, but your friends won't call you Einstein unless you let your hair grow, parade around in a messy old sweater, and stick out your tongue when they take your picture.

Space: A Commotion of Motion

Everything in space is moving and turning. Objects can't sit still. Thanks to gravity, other celestial bodies are always pulling on a star, planet, galaxy, or spacecraft. Some of us are self-centered, but the universe has no center.

For example, Earth

- Turns on its axis — what astronomers call *rotating* — and takes one day to turn all the way around.

- Orbits around the Sun — what astronomers call *revolving* — with one complete orbit taking one year.

- Travels with the Sun in a huge orbit around the center of the Milky Way. The trip takes about 226 million years to complete once, and the duration of the trip is called the *galactic year*.

- Moves with the Milky Way in a trajectory around the center of the *Local Group of Galaxies,* a couple dozen galaxies in our neck of the universe.

- Moves through the universe with the Local Group as part of the *Hubble Flow,* the general expansion of space caused by the Big Bang.

  The *Big Bang* is the event that gave rise to the universe and set space itself expanding at a furious rate. Detailed theories about the Big Bang explain many observed phenomena and have successfully predicted some that hadn't been observed before the theories were circulated. (For more about the Big Bang and other aspects of the universe, check out Part IV.)

Remember Ginger Rogers? She did everything Fred Astaire did when they danced in the movies, and she did it all backward. Like Ginger and Fred, the Moon follows all the motions of Earth (although not backward), except for Earth's rotation; the Moon rotates more slowly, about once a month. And it performs its tasks while also revolving around Earth (which it does about once a month).

And you, as a person on Earth, participate in the motions of rotation, revolution, galactic orbiting, Local Group cruising, and cosmic expansion. You do all that while you drive to work, whether you know it or not. Ask your boss for a little consideration the next time you run a few minutes late.

Chapter 2

Join the Crowd: Skywatching Activities and Resources

*A*stronomy has universal appeal. The stars have fascinated people everywhere from prehistoric times into the modern age. Early observations of the sky led to all sorts of theories about the universe and attributions of power and purpose to the movements of stars, planets, and comets. As you look up at the sky, hundreds of thousands of people worldwide watch with you. When it comes to skywatching, you're not alone. Many people, organizations, publications, websites, smartphone apps, and other resources are at your disposal to help you get started, to keep you going, and to help you participate in the great work of explaining the universe.

In this chapter, I introduce you to these resources and give you suggestions on how you can get started. The rest is up to you. So join in!

When you know the resources, organizations, facilities, and equipment that can help you enjoy astronomy more deeply, you can move comfortably to the science of astronomy itself — the nature of the objects and phenomena out there in deep space. I describe the equipment you need to get started in Chapter 3.

You're Not Alone: Astronomy Clubs, Websites, Smartphone Apps, and More

You have plenty of readily available information, organizations, people, and facilities to help you get started and remain active in astronomy. You can join associations and activities to help researchers keep track of stars and planets. You can attend astronomy club meetings, lectures, and instructional sessions, which allow you to share telescopes and viewing sites to enjoy the sky with others. You also can find magazines, websites, books, computer programs, and smartphone apps with basic information on astronomy and current events in the sky.

Depending on where you live, you may even have access to TV programs that keep you in touch with the cosmos, like the BBC's *The Sky at Night,* in Britain, and the Public Broadcasting System's *SkyWeek,* in the United States.

Joining an astronomy club for star-studded company

The best way to break into astronomy without undue effort and expense is to join an astronomy club and meet the regulars. Clubs hold monthly meetings where the old hands pass along tips on techniques and equipment to beginners, and where local and visiting scientists present talks and slide shows. Members likely know where to get a good deal on a used telescope or binoculars and which products on the market are worth the money. (See Chapter 3 for more.)

Even better, astronomy clubs sponsor observing meetings, usually on weekend nights and occasionally on special dates during a meteor shower, an eclipse, or another special event. An observing meeting is the best place to find out about the practice of astronomy and the equipment you need. You don't even need to bring a telescope; most folks are happy to give you a look through theirs. Just wear sensible shoes, bring mittens and a hat for the cool night air, and put on a smile!

If you live in a city or a suburb, chances are good that your night sky is bright, so you can find better observational conditions if you travel to a dark spot in the country. Your local astronomy club probably already has a good spot, and when the members converge on that lonely place, you can enjoy safety in numbers.

Gazing around the world: A sampling of astronomy clubs

The Astronomical Society of the Pacific (www.astrosociety.org), with headquarters in San Francisco, publishes the quarterly digital magazine *Mercury* for amateurs. It holds an annual meeting that moves around the western United States and sometimes goes as far east as Boston or Toronto. The society also offers numerous educational materials in astronomy to teachers.

Do you live in Canada? The Royal Astronomical Society of Canada has 29 *Centres,* which is a fancy name for astronomy clubs. Professionals from the nearest university often get involved with the Centre's activities. To find a Centre near you, check out the map of Canada on the RASC website, www.rasc.ca. Click a city name to access the Centre website for that location.

In the United Kingdom, the venerable British Astronomical Association, founded in 1890, is still going strong. Its website is www.britastro.org. And the Society for Popular Astronomy, billing itself as "Britain's brightest astronomical Society," features sky event news and tweets about planets, bright meteors, and more on its colorful website, www.popastro.com.

Most other countries have astronomy clubs, too. Astronomy is truly a "universal" passion.

If you live in a good-size city or a college town, you can probably find an astronomy club nearby. If you live in the United States, find the club(s) near you with the locator form of the NASA Night Sky Network, at http://nightsky.jpl.nasa.gov. You can also check out the website of America's "club of clubs," the Astronomical League, at www.astroleague.org. Browse the list of more than 240 member societies, arranged by state.

For a more global approach, visit the *Sky & Telescope* website, at www.skyandtelescope.com (just click the "Clubs and Organizations" tab in the Community menu to find clubs and organizations worldwide). Insert your city, state or province, and country to see information on the nearest amateur astronomy group(s). In Kansas City, Missouri, for example, skywatcher wannabes have four groups to choose from. The site lists five organizations for London, England, and in Vancouver, British Columbia, you can join a club with more than 300 members.

Checking websites, magazines, software, and apps

Finding out about astronomy is easy. You can choose from a wide range of resources, including websites, apps for smartphones and tablet computers, magazines, and desktop computer software. The following sections offer some tips for finding the best information.

Traveling through cyberspace

The Net offers sites on every topic in astronomy, and the resources are increasing at, well, an astronomical rate! You can find many websites listed throughout this book; if you want more information on planets, comets, meteors, or eclipses, the web offers good sites on every topic.

The editors of *Sky & Telescope* magazine maintain one of the best websites, at www.skyandtelescope.com. Get your observational career started by checking out *Sky & Telescope*'s "This Week's Sky at a Glance" page, at www.skyandtelescope.com/observing/ataglance. It gives a day-by-day (or night-by-night) account of planets, comets, and other sights to look for.

Outside the United States, you can probably find a website tailored to your country, or at least your latitude north or south of the equator. In the United Kingdom, for example, *Astronomy Now* magazine offers an Interactive Star Chart designed for observers in Great Britain and Ireland, at www.astronomynow.com/sky_chart.shtml.

Perusing publications

You can purchase excellent magazines to expand your knowledge of astronomy and your skill at practicing it. Most amateur astronomers subscribe to at least one publication. In many cases, if you join a local astronomy club, you may have access to a subscription to a national magazine at a member discount. (See "Joining an astronomy club for star-studded company," earlier in this chapter, for the scoop on clubs.)

I recommend that you pick up a copy of each of the "big two" (literally, the biggest two) astronomical magazines: *Sky & Telescope* and *Astronomy*. Test-drive the publications for a month, and if you get more out of one than the other, go ahead and subscribe. You can do so from their websites at www.skyandtelescope.com and www.astronomy.com. (You can use the observing aids and other helpful resources on the *Sky & Telescope* site whether or not you subscribe to the magazine, but much of the material on the *Astronomy* site is for subscribers only.)

Canadian readers can get the bimonthly *SkyNews: The Canadian Magazine of Astronomy & Stargazing,* a slick, full-color publication. Subscribers become Astronomy Associates of the Canada Science and Technology Museum. Call 1-866-759-0005 or visit www.skynews.ca for information.

Astronomy buffs in the United Kingdom should look for *Astronomy Now* and *Popular Astronomy* to see which magazine they prefer. Check their websites at www.astronomynow.com and www.popastro.com/popularastronomy.

In France, an excellent and well-illustrated magazine is *Ciel & Espace* (www.cieletespace.fr); in Australia, look for *Australian Sky & Telescope* (www.austskyandtel.com.au); and in Germany, *Sterne und Weltraum* (www.astronomie-heute.de) excels.

Wherever you live, you need the annual *Observer's Handbook* of the Royal Astronomical Society of Canada (www.rasc.ca). Dozens of experts compile the handbook to help you enjoy the skies.

Surveying software and apps

A planetarium program, or "desktop planetarium," for your personal computer is a real plus. So is a planetarium app for your smartphone or tablet computer. Such programs or apps show you what the sky looks like from your home every night. You can also use them to find out what stars and planets will be up in the sky at a future date or at a different location so you can check in advance what you may observe on an upcoming vacation or visit to a dark sky observation site. This software is terrific to look at before you step outside to view the night sky. Some astronomers use these programs to plan their observing sessions. They prepare schedules of objects to scan with telescopes and binoculars at different times of the night to use their "dark time" effectively. Amateurs with certain telescope models that feature computer control can use some planetarium programs to guide their telescopes to stars, planets, or other sky objects of interest.

Desktop planetarium programs are available over a wide price range (including free programs) with many different features. You can find some programs advertised in astronomy and science magazines and on websites (see the previous two sections); they're updated occasionally for increased usefulness. You need only one program to get started, and that one may be the only program you ever need. The best way to select the planetarium program that suits you is to talk to experienced amateur astronomers at your local astronomy club. What works for them likely will work for you.

Here are a few programs that I think are worthy of your consideration:

- ✔ **Stellarium:** A free, open-source personal planetarium program for use on computers running most operating systems. Visit the Stellarium website at `www.stellarium.org` to learn about its many features, view sample screen shots, or download the program to your computer.

- ✔ **TheSkyX:** Three versions of the TheSkyX planetarium program are available from Software Bisque, at `www.bisque.com/sc`, but I recommend the simplest and cheapest version, TheSkyX Student, for beginners. You can use TheSkyX Student to display what the sky looks like at any time and place on Earth, and it even shows what stars are up in the daytime (although you can't see them except during a total eclipse of the Sun). You can print star charts and even use it to predict Iridium satellite flares visible from your location. (I describe Iridium flares in Chapter 4.) But this simple program won't point your telescope for you.

- ✔ **Sky Safari Pro:** Available only for Mac computers, this program comes in three versions. The basic version doesn't control a telescope but probably has anything else that a beginning amateur astronomer needs. This program is based on the Sky Safari app, which I describe later in this section. It's available from `www.southernstars.com`.

- ✔ *Sky and Telescope***'s online planetarium program:** You can find a simple (and free) online planetarium program at `www.skyandtelescope.com`. An interactive sky chart comes up on your screen, showing the sky over Greenwich, England, at the present time. You can reset it for your own geographic location and for whatever time and date you plan to go stargazing.

A great many astronomy-related apps for smartphones and tablet computers have appeared on the market in the last few years. I have space here to just list a few of them:

- ✔ **AstroGizmo:** This inexpensive planetarium program, for the iPhone and iPad, shows the 88 constellations (which I list in Chapter 1) and the planets. Just hold your device up to the sky, and AstroGizmo shows you the stars and the planets (if any) in that direction, as seen from your location at the current time.

- ✔ **CraterSizeXL:** Use this iPad and iPhone app to calculate the possible danger if a potentially hazardous asteroid is headed for Earth (I discuss PHAs in Chapter 7). Fill in the available information on the object, and CraterSizeXL predicts the impact energy in units of Hiroshima-equivalent atom bomb blasts, crater size, and more. Damage from an asteroid hit can be in the trillions of dollars, but the good news is that you can download the app for about a buck.

- ✔ **Distant Suns 3:** Yet another app for iPhone and iPad, this planetarium program is particularly appropriate for American amateur astronomers because it alerts you to upcoming astronomy club events in the United States with information from the NASA Night Sky Network.

- ✓ **Galaxy Zoo:** For Android phones and tablets, and for iPhone and iPad, this free app is for citizen scientists who help advance the science of astronomy by classifying the shapes of an astronomical number of galaxies that the Hubble Space Telescope photographed. Join more than a quarter-million volunteers worldwide in this worthy effort. (I describe galaxies and how to join Galaxy Zoo with your home computer, smartphone, or tablet in Chapter 12.)

- ✓ **Google Sky Map:** If you have an Android phone or tablet, you can use this free app to identify visible stars and planets or to enjoy images of numerous celestial objects from NASA and other sources.

- ✓ **GoSatWatch:** iPhone and iPad owners can use this app to learn where artificial satellites are orbiting and to predict when satellites will pass over your location (or any other) on Earth. (I describe artificial satellite observing in Chapter 4.)

- ✓ **NASA Meteor Counter:** This app for the iPhone and iPad helps you count and report meteors to NASA, as I describe in Chapter 4.

- ✓ **Sky Safari:** This highly rated planetarium program (for Android devices, iPhone, and iPad) is available in several versions at prices ranging from about $3 for the simplest version to about $60 for the most advanced. The more you pay, the more features you get. With the basic version, you can point the phone toward the sky, night or day, and it identifies the celestial objects visible (night) or invisible (day) that are up in that direction. Start with the cheap version and see if it does everything you need.

- ✓ **Space Junk Lite:** Android phone and tablet owners can use this free app to find out what stars, planets, and constellations are visible in the night sky at their locations and spot the International Space Station and Hubble Space Telescope as they cross the night sky. If you like this app, consider downloading the advanced version, Space Junk Pro (about $5).

- ✓ **Stellarium:** This free iPhone app is based on the free Stellarium desktop planetarium program that I describe earlier in this section.

Visiting Observatories and Planetariums

You can visit *professional observatories* (organizations that have large telescopes staffed by astronomers and other scientists for use in studying the universe) and *public planetariums* (specially equipped facilities with machines that project stars and other sky objects in a darkened room, accompanied by simple explanations of sky phenomena) to find out more about telescopes, astronomy, and research programs.

Ogling the observatories

You can find dozens of professional observatories in the United States and many more abroad. Some serve as research institutions operated by colleges and universities or government agencies. Examples include the U.S. Naval Observatory (located in the heart of Washington, D.C., guarded by the Secret Service, and home to the Vice President of the United States; www. usno.navy.mil/USNO) and outposts on remote mountaintops (such as the University of Denver's Mt. Evans Meyer-Womble Observatory, billed as the "Highest Operating Observatory in the West," at 14,148 feet; http:// mysite.du.edu/~rstencel/MtEvans).

Some observatories are dedicated to public education and information; cities, counties, school systems, or nonprofit organizations often operate these facilities. Here are some of the top sites:

- **The Royal Observatory, Greenwich, in London, England:** One of the most famous observatories in the world, at one time the Royal Observatory (www.rmg.co.uk/royal-observatory) was a professional research facility, then called the Royal Greenwich Observatory. The observatory is "home" of the Greenwich meridian, from which longitude is measured around Earth. Equally important, it was the original source of Greenwich Mean Time, which set the standard for timekeeping worldwide.

- **Lowell Observatory, on Mars Hill in Flagstaff, Arizona:** Research observatories vary in the degree to which they accommodate visits from the public, but Lowell is especially welcoming. You can even look at planets or stars on some nights: The observatory advertises that you can "Peer through the telescope that Percival Lowell used to sketch Mars or visit the telescope that helped Clyde Tombaugh discover Pluto." This observatory has a fine visitor center with a theater and exhibit hall and offers frequent tours. Check out its website at www.lowell.edu. (I discuss Percival Lowell and Mars in Chapter 6 and Pluto in Chapter 9.)

- **The National Solar Observatory, at Sunspot, New Mexico:** This observatory runs Sun-watching telescopes above the little town of Cloudcroft, which is high above the city of Alamogordo. You can check out its Sunspot Astronomy and Visitor Center (daytime visits only) and take a tour of the observatory as well. See www.nso.edu for more.

- **Mount Wilson Observatory, in the San Bernardino Mountains above Los Angeles, California:** Mount Wilson, where the expansion of the universe and the magnetism of the Sun were discovered, is a landmark in the history of science. Albert Einstein was a special guest there, but you don't have to be a Big Brain to enjoy your visit. See www.mtwilson.edu.

- **The Griffith Observatory and Planetarium, in Los Angeles, California:** This observatory is operated entirely for the public at Griffith Park in Los Angeles and is well worth visiting. Check out www.griffithobs.org.

- **Palomar Observatory, near San Diego, California:** Here you can see the famous 200-inch telescope that, for decades, was the largest and best in the world. Even with newer instrumentation, the telescope is still a great contributor of new knowledge on the universe. You have to guide yourself around the observatory grounds. Check the website, at `www.astro.caltech.edu/palomar`, just before you visit; the facility closes well before sunset and is often off limits on short notice due to road and weather conditions near the summit. The observatory has a small museum and gift shop.

- **Kitt Peak National Observatory, on a Native American reservation (the Tohono O'odham Nation) in the Sonoran desert, 56 miles west of Tucson, Arizona:** When I worked at Kitt Peak during the 1960s, tourists were permitted to visit only during the day. Even so, there was a lot to see, from the visitor center (an astronomical museum) to the many telescope domes. Things have changed now, so put KPNO high on your list when you visit the American Southwest. During the day, you can take part in a guided or self-guided tour. You can also experience nighttime observing with certain KPNO telescopes. You must make an appointment for observing in advance. For details, go to `www.noao.edu/kpno` and click "Tourist Information."

- **MMT Observatory, on Mount Hopkins in the Coronado National Forest, 37 miles south of Tucson, Arizona:** Check in at the Fred Lawrence Whipple Visitor Center at the base of the mountain, where you can enjoy exhibits and sign up for tours of the observatory, with its 256-inch (6.5-meter) reflecting telescope, one of the largest in the continental United States. But before you go, check the MMT public tours website at `www.mmto.org/node/289` to find when the visitor center is open and when tours are offered.

- **Mauna Kea Observatories, on the Big Island of Hawaii:** The biggest "telescope farm" in the United States is home to 13 large telescopes operated by the United States and ten other nations. No comparable assemblage of big telescopes exists in the Northern Hemisphere; in the Southern Hemisphere, the only comparable assemblage is the European Southern Observatory in Chile.

  Mauna Kea, is well worth visiting, but you need to be in good health, due to the high altitude (13,796 feet), and follow the visitor instructions at the MKO website (`www.ifa.hawaii.edu/mko`). The first time I visited the observatory, I walked up one flight of stairs and began to turn blue. My hosts administered oxygen and drove me down to a staff facility at a "mere" 9,000 feet to recover. There's also a visitor information station at 9,000 feet, where you can start your visit.

You can also visit radio astronomy observatories, where scientists "listen" to radio signals from the stars or even seek signals from alien civilizations. Here are my top picks:

✔ **The National Radio Astronomy Observatory:** Sponsored by the National Science Foundation, the NRAO has facilities near Socorro, New Mexico, and Green Bank, West Virginia, that welcome visitors (see www.nrao.edu):

- **Very Large Array (VLA) Visitor Center:** Drive across the Plains of San Agustin, near Socorro, to see the huge radio telescope, a system of 27 dish-shape radio telescopes, each 82 feet in diameter. Check out the visitor center and gift shop, but if you want a tour, you have to come on the first Saturday of the month. Tour times and other details are at www.nrao.edu/index.php/learn/vlavc.

- **Robert C. Byrd Green Bank Telescope:** At Green Bank, nestled among mountains in the United States National Radio Quiet Zone, you can tour the 328-foot telescope, which is the world's largest fully steerable dish antenna. The Green Bank Science Center offers interactive exhibits. After taking in the sights, stop by the Starlight Café to take in some tasty refreshments. Check out the visiting hours at www.nrao.edu/index.php/learn/gbsc.

✔ **Jodrell Bank Observatory, near Goostrey, Cheshire, England:** At this observatory, operated by the University of Manchester, you can see the historic 250-foot Lovell Telescope, a big dish once used to bounce radar signals off Soviet booster rockets and even off the Moon. The Jodrell Bank Discovery Centre offers a Space Pavilion, Planet Walk, Galaxy Garden, and more. It's closed on some holidays, so check the website (www.jodrellbank.net/visit) before you go.

✔ **Parkes Radio Telescope, near Parkes, New South Wales, Australia:** The radio telescope to visit when you are Down Under is the 210-foot Parkes Radio Telescope. The telescope is known to astronomers for its research findings, but it achieved its greatest public notice when it relayed radio transmissions to NASA from Apollo astronauts on their missions to the Moon. The Parkes Radio Telescope Visitors Discovery Centre has exhibits, two theaters, and the aptly named Dish Café. See the website at www.csiro.au/Portals/Education/Programs/Parkes-Radio-Telescope/VisitParkesTelescope.aspx.

Popping in on planetariums

Planetariums, also called *planetaria,* are just right for beginning astronomers. They provide instructive exhibits and project wonderful sky shows indoors on the planetarium dome or on a huge screen. And many offer nighttime sky-watching sessions with small telescopes, usually held outside in the parking lot, in an adjacent small observatory dome, or at a nearby public park. Many have excellent shops where you can browse the latest astronomy books, magazines, and star charts. The planetarium staff can also direct you to the nearest astronomy club, which may even meet after hours in the planetarium itself.

I practically grew up in the Hayden Planetarium at the American Museum of Natural History in New York City. Occasionally, I confess, I even snuck in for free. The planetarium staff was nice enough to have me back to speak (also for free) at its 50th anniversary. Although the old planetarium was torn down, a spectacular new one has replaced it. Make this planetarium a prime destination next time you visit the Big Apple. It's pricey but still much cheaper than a Broadway show, and its stars never miss a cue or sing off key (just don't try to sneak in like me!). Visitor information is on the Hayden's website, at www.amnh.org/visitors/index.php.

 Loch Ness Productions, in Colorado, is the monster of the planetarium business. It keeps tabs on more than 3,500 planetariums and domed theaters worldwide. Loch Ness doesn't list them all online, but you can probably find a planetarium near you, or in a country that you plan to visit, in its online directory, at www.lochnessproductions.com/lfco/lfco.html.

Vacationing with the Stars: Star Parties, Eclipse Trips, Dark Sky Parks, and More

An astronomy vacation is a treat for the mind and a feast for the eyes. Plus, traveling with the stars is often cheaper than taking a conventional holiday. You don't have to visit the hottest tourist destinations to keep up with your snooty neighbors. You can have the experience of a lifetime and come back raving about what you saw and did, not just what you ate and spent.

However, you can blow big bucks on one type of astronomy vacation: the eclipse cruise. But if you like ocean cruises, taking one to an eclipse doesn't cost any more than making a similar voyage that has no celestial rewards. Bargain-basement eclipse tours are available, too. Star parties, telescope motels, and visits to dark sky parks are additional options; I cover all the bases in the following sections. Pack your bags and have the neighbors watch the dog!

Party on! Attending star parties

Star parties are outdoor conventions for amateur astronomers. They set up their telescopes (some homemade and some not) in a field, and people take turns skywatching. (Be prepared to hear plenty of "Oohs" and "Ahs.") Judges choose the best homemade telescopes and equipment, earning their owners esteem and sometimes even a prize. If rain falls in the evening, partygoers may watch slide shows in a nearby hall or a big tent. Arrangements vary, but often some attendees camp in the field; others rent inexpensive cabins or commute from nearby motels. Star parties may last for a night or two, or sometimes as long as a week. They attract a few hundred to a few thousand (yes, thousand!) telescope makers and amateur astronomers. And the larger star parties have websites with photos of previous events and details on coming attractions. Some resemble the AstroFests I mention later in this section, with exhibitors and distinguished speakers as well as stargazing.

The leading star parties in the United States include

- **Stellafane:** This Vermont star party has been going strong since 1926 (http://stellafane.org/convention/index.html).
- **Texas Star Party:** Commune with the stars on the mile-high Prude Ranch in the Lone Star State (http://texasstarparty.org).
- **RTMC Astronomy Expo:** You can get high on the stars at an elevation of 7,600 feet in California's San Bernardino Mountains (www.rtmcastronomy expo.org).
- **Enchanted Skies Star Party:** Head to the desert in New Mexico, with dark sky observing on South Baldy, at an altitude of 10,600 feet (http://enchantedskies.org).
- **Nebraska Star Party:** This party boasts "a fantastic light pollution–free sweep of the summer night sky" (www.nebraskastarparty.org).

Here are some of the leading star parties in the United Kingdom:

- **The Equinox Sky Camp:** Held at Kelling Heath, Norfolk, this party bills itself as "the largest star party in the U.K." (www.starparty.org.uk).
- **Kielder Forest Star Camp:** Located in Northumberland, Kielder Forest is thought to be "the darkest venue for any English star party" (www.kielderforeststarcamp.org/starcamp.html).

If you live in or plan to visit the Southern Hemisphere, check out this star party:

- **South Pacific Star Party:** It's held near Ilford, NSW, Australia, on a property reserved for skywatching by the Astronomical Society of New South Wales (www.asnsw.com/node/712).

In the long run, I recommend that you visit at least one of these star parties, but in the meantime, you can ask at a local astronomy club meeting about a similar event that may be planned in your vicinity.

Getting festive at an AstroFest

You can hear famed astronomers, meet science authors, and learn the latest space news at an expo for astronomy enthusiasts. Astronomy groups organize the expos to reach out to the public, students, and educators. The events, often called AstroFests, feature talks on astronomy and space research, as well as displays and demonstrations of the latest technology available to amateur astronomers. The expos are most common under the AstroFest name in Europe and Australia. Many of the large star parties (which I describe in the preceding section) in the United States also feature daytime speakers and exhibitors, as well as nighttime observing.

To the path of totality: Taking eclipse cruises and tours

Eclipse cruises and tours are planned voyages to the places where you can view total eclipses of the Sun. Astronomers can calculate long in advance when and where an eclipse will be visible. The locations where you can see the total eclipse are limited to a narrow strip across land and sea, the *path of totality.* You can stay home and wait for a total eclipse to come to you, but you may not live long enough to see more than one, if any. So if you're the impatient stargazing type, you may want to travel to the path of totality.

Recognizing reasons to book a tour

If an eclipse is visible within easy driving distance, you don't need to sign up for a tour. (But such cases are rare; see the list of upcoming total eclipses of the Sun in Chapter 10, Table 10-1.)

If you're an experienced domestic and international traveler, you can go on your own to the path of totality for a distant eclipse. But consider this fact: Expert meteorologists and astronomers identify the best viewing locations years in advance. More often than not, these places aren't vast metropolises that offer huge numbers of vacant accommodations. You have to travel to random spots on the globe. After experts tab a spot as a prime location for a coming eclipse, tour promoters and savvy individuals book all or most of the local hotels and other facilities years in advance. Johnnies-or-Janies-come-lately, especially those who travel on their own, may be out of luck.

A tour promoter usually engages a meteorologist and a few professional astronomers (sometimes even me!). So you have the benefit of a weatherperson to make last-minute decisions on moving the group's observing site to a place with a better next-day forecast, an astronomer to show you the safest methods to photograph the eclipse, and usually another lecturer who tells old eclipse tales and reports on the latest discoveries about the Sun and space.

On the night following the eclipse, everyone shows their videos of the darkening sky, the birds coming to roost in the middle of the day, a klutz knocking over his telescope at the worst possible time, and the excited crowd saying "Wow" and "Hurray." And, of course, they replay the eclipse — over and over.

If these details haven't convinced you to take an eclipse tour for your viewing pleasure, consider this: Taking a group tour to a foreign destination is often cheaper than going on your own (and more satisfying than waiting years until a total eclipse is visible near your home). In the group, you'll make new friends who share your passion for eclipse chasing and skywatching. It's certainly happened to me.

Examining the advantages of cruising

An eclipse cruise is usually much better than a tour but more expensive. At sea, the captain and navigator have "2° of freedom." When the meteorologist says, "Head southwest down the path of totality for 200 miles" on the night before the eclipse (because of a best-case forecast for a cloud-free location at eclipse time), the ship can follow those instructions. But on land, you have to keep the bus on the road, and a road may not go in the direction you need it to go. At one total eclipse, in Libya, our procession of tour busses headed off the nearest road, proceeding across the desert to the designated viewing site, where water, port-a-pottys, security guards, and T-shirt vendors were on hand. On a cruise, you can leave the steering to the crew, recline in a deck chair, sip a piña colada, have your camera ready, and wait for totality.

I've viewed many eclipses, and my experience is that if you stay on land, you get a clear view of totality about half the time. But if you travel on an ocean liner, you almost never miss.

Making the right decision

You can find advertisements for eclipse tours and cruises in astronomy and nature magazines, on astronomy and travel websites, and in many science and nature magazines. Astronomy clubs, fraternal organizations, and college alumni societies often organize group accommodations on eclipse cruises.

Here are some ways you can choose the right tour or cruise for you:

- ✔ **Consult current and back issues of astronomy magazines.** Most run articles about the viewing prospects for a solar eclipse a few years in advance. Get their expert recommendations about the best viewing sites.

- ✔ **Check out the advertisements from travel operators.** Which tours and cruises go to the best places? Get brochures from travel agents, tour promoters, and cruise lines. Promoters often list previous successful eclipse trips, indicating a level of experience.

You can do it! Participating in scientific research

You can make your astronomy hobby beneficial and fun by joining national and worldwide efforts to gather precious scientific data. You may have only a pair of binoculars compared to Keck Observatory's two 10-meter-wide (400-inch) telescopes, but if the observatory has a cloudy day, Keck can't see anything. And if a spectacular fireball shoots over your hometown, you may be the only astronomer to see it.

Secret U.S. Department of Defense satellites and an amateur moviemaker vacationing at Glacier National Park recorded one of the most spectacular and interesting meteors of all time. A clip from that home movie appears in just about every scientific documentary about meteors, asteroids, and comets that appears on television. It pays to be in the right place

at the right time. And someday you may be in that position.

Join other amateur astronomers in so-called citizen science and enjoy the projects I recommend throughout this book. You can use your smartphone to help NASA's meteor count (see Chapter 4), search through photos made by a Moon-orbiting satellite to identify small lunar craters (see Chapter 5), look through radio data from the Allen Telescope Array to search for extraterrestrial intelligence (see Chapter 14), help scientists find planets of stars beyond the Sun (also see Chapter 14), or search for bubble-shape nebulae in the Milky Way and classify galaxies photographed by the Hubble Space Telescope on your home computer (both projects are described in Chapter 12).

Motoring to telescope motels

Telescope motels are establishments where the attractions are the dark skies and the opportunity to set up your own telescope in an excellent viewing location. They usually have telescopes of their own that you can use, perhaps at an additional fee. If you fancy enjoying a stargazing vacation without lugging your own equipment across the country or the world, telescope motels are a good option.

Check out these telescope motels worth visiting in the United States:

- ✔ **The Observer's Inn:** Located in the historic gold-mining town of Julian, California, this hotel has an observatory and concrete pads on which guests can place their own telescopes. Check it out online at www. observersinn.com.

- ✔ **Primland:** Situated in the Blue Ridge Mountains near Meadows of Dan, Virginia, Primland is a luxury resort, no mere motel; it features stargazing in its own observatory (see http://primland.com). Invite your rich uncle and suggest that it's his treat.

✔ **San Pedro Valley Observatory:** Located near Benson, Arizona, San Pedro Valley Observatory provides astronomers/guides with a 20-inch telescope and several smaller ones for observations and astrophotography. See http://arizona-observatory.com for more details. It doesn't offer lodging, but instead refers you to accommodations at nearby motels and bed-and-breakfasts.

Overseas, telescope motels worth visiting include:

✔ **Fieldview Guest House:** This bed-and-breakfast has six telescopes and is located in Norfolk, in the East of England. See www.fieldview.net for more info.

✔ **COAA (Centro de Observação Astronómica no Algarve):** Located in southern Portugal, this facility features several guest suites, a half-dozen telescopes, and a radar system to observe meteors day and night. Check it out online at www.coaa.co.uk/index.html.

✔ **SPACE (San Pedro de Atacama Celestial Explorations):** This telescope motel in San Pedro de Atacama, Chile, offers lodging, tours (guides speak English, Spanish, and French), and telescopes on the high Atacama desert at an elevation near 8,000 feet. (See www.spaceobs. com.) The region is considered one of the finest locations for astronomical observatories on Earth.

Setting up camp at dark sky parks

The International Dark-Sky Association (IDA, www.darksky.org) confers the designation of International Dark Sky Park on public lands with good starry skies and limited or very low interference from artificial lighting. A dark sky park may or may not have telescopes of its own, but it's a place you can visit for a good view of the sky and to set up your own portable telescope.

Here are dark sky parks worth your visit in the United States:

✔ **Natural Bridges National Monument:** This dark sky park in Utah claims that at night a bridge "forms a window into a sky filled with thousands of stars bright enough to cast a shadow." For more, see www.nps.gov/nabr/index.htm.

✔ **Big Bend National Park:** This dark sky park is in Texas, on the Rio Grande. You can find it online at www.nps.gov/bibe/index.htm.

✔ **Geauga County Observatory Park:** This Ohio park has telescopes and weather and seismic stations. Its website is www.geaugaparkdistrict.org/observatorypark.shtml.

- ✔ **Cherry Springs State Park:** The Susquehannock State Forest in Pennsylvania lays claim to this dark sky park, which often hosts star parties. Find out more at www.dcnr.state.pa.us/stateparks/findapark/cherrysprings/index.htm.

- ✔ **Clayton Lake State Park:** Watch the New Mexico night sky from the public observatory or listen to sky talks by astronomers. See www.emnrd.state.nm.us/PRD/Clayton.htm.

- ✔ **Goldendale Observatory State Park:** This dark sky park in Washington State features telescopes for nighttime use, "constellation tours," and daytime views of Venus when the planet is in the afternoon sky. Find it online at www.perr.com/gosp.html.

- ✔ **The Headlands:** View the night sky along the Straits of Mackinac, in northern Michigan. The website www.emmetcounty.org/headlands has more information.

In Europe, consider exploring these dark sky parks:

- ✔ **Galloway Forest Park:** This dark sky park is located in Scotland. See www.forestry.gov.uk/darkskygalloway.

- ✔ **Sark Dark Sky Island:** In the Channel Islands off the coast of Normandy, Sark Dark Sky Island boasts "the only light pollution here is a distant glow from Guernsey, Jersey, and France." Check out its website, www.socsercq.sark.gg/darkskies.html.

- ✔ **Exmoor National Park:** Located in the southwest of England, Exmoor was the first designated International Dark Sky Reserve in Europe. See www.exmoor-nationalpark.gov.uk.

- ✔ **Hortobágy National Park:** This dark sky park is in Hungary. Its website is www.hnp.hu/index_en.php.

In the Southern Hemisphere, you can visit:

- ✔ **Aoraki Mackenzie International Dark Sky Reserve:** This huge reserve on New Zealand's South Island extends over more than 1,600 square miles. Read more about it at the IDA website www.darksky.org/IDSReserves.

When you enjoy an international dark sky park, you'll recognize that "lights out" is a good thing.

Chapter 3

The Way You Watch Tonight: Terrific Tools for Observing the Skies

If you've ever gone outside and looked at the night sky, you've been star-gazing — observing the stars and other objects in the sky. Naked-eye observation can distinguish colors and the relationships between objects — like finding the North Star by using "pointer stars" in the Big Dipper.

From naked-eye observation, you can take a short step up by adding optics to see fainter stars and view objects with greater detail. First try binoculars, and then graduate to a telescope. Next thing you know, you're an astronomer!

But I'm getting ahead of myself. First, you need to take some quiet looks at the cosmos and see the beauty and mystery for yourself. You can use three basic tools, at least one of which you already own.

Whether you use your eyes, a pair of binoculars, or a telescope, each method of observation is best for some purposes:

✔ **The human eye:** Your peepers are ideal for watching meteors, the aurora borealis, a planetary conjunction (when two or more planets are close to each other in the sky), or a conjunction of a planet and the Moon.

✔ **Binoculars:** A good pair of binoculars is best for observing bright variable stars, which are too far from their *comparison stars* (stars of known constant brightness used as a reference to estimate the brightness of a star that varies in brightness) to see them together through a telescope. And binoculars are wonderful for sweeping through the Milky Way and viewing the bright nebulae and star clusters that dot it here and there. Some of the brighter galaxies — such as M31 in Andromeda, the Magellanic Clouds, and M33 in Triangulum — look best through binoculars.

✔ **Telescope:** You need a telescope to get a decent look at most galaxies and to distinguish the members of close double stars, among many other uses. (A double star consists of two stars that appear very close together; they may or may not be near each other in space, but when they truly are together, they form a binary star system.)

In this chapter, I cover these observational tools, provide you with a quick primer on the geography of the night sky, and give you a handy plan for delving into astronomy. Before long, you'll be observing the skies with ease.

Seeing Stars: A Sky Geography Primer

When viewed from the Northern Hemisphere, the whole sky seems to revolve around the North Celestial Pole (NCP). Close to the NCP is the North Star (also called Polaris), a good reference point for stargazers because it always appears in almost exactly the same place in the sky, all night long (and all day long, but you can't see it then).

In the following sections, I show you how to familiarize yourself with the North Star and give you a few facts about constellations.

As Earth turns . . .

Our Earth turns. The Greek philosopher Heraclides Ponticus proclaimed that concept in the fourth century B.C. But people doubted Heraclides' observations because folks thought that, if his theories were true, they should feel dizzy like riders of a fast merry-go-round or a whirling chariot. They couldn't imagine a turning Earth if they didn't physically feel the effects. Instead, the ancients thought the Sun raced around Earth, making a complete revolution every day. (They didn't feel the effects of Earth's rotation, nor do you and I, because those effects are too small to notice.)

Proof of Earth's turning, or *rotation,* didn't come until 1851, more than two millennia after Heraclides (researchers didn't have much government funding back then, so progress was slow). The proof came from a big French swinger: a heavy metal ball suspended from the ceiling of a church in Paris (the Panthéon) on a 220-foot wire. The device is called a *Foucault pendulum,* after the French physicist who came up with the plan. If you kept an eye on the pendulum as it swung back and forth all day, you could see that the direction taken by the swinging ball across the floor gradually changed, as though the floor was turning underneath it. And it was — the floor turned with Earth.

If you're not convinced that Earth turns, or if you just like to watch swingers, you can see a 240-pound brass Foucault pendulum in the rotunda of the Griffith Observatory, in Los Angeles (`www.griffithobs.org/exhibits/brotunda.html`). In England, check out the University of Manchester's Foucault pendulum (`www.mace.manchester.ac.uk/project/teaching/civil/structural concepts/Dynamics/pendulum/pendulum_pra3.php`).

If you already believe that Earth turns, however, you can verify that conclusion as you just sip a favorite beverage and watch the Sun set in the west.

As I explain in Chapter 1, the rotation of Earth around its axis makes the stars and other sky objects appear to move across the sky from east to west. In addition, the Sun moves across the sky during the year on a circle called the *ecliptic.* (If you could see the stars in the daytime, you'd note that the Sun moving to the west across the constellations, day by day.) The ecliptic is inclined by 23.5° to the celestial equator, the same angle by which the axis of Earth is tilted from the perpendicular to its orbital plane.

The planets stay close to the ecliptic as they move throughout the year. They move systematically through 12 constellations located on the ecliptic, which, collectively, are called the *Zodiac:* Aries, Taurus, Gemini, Cancer, Leo, Virgo, Libra, Scorpius, Sagittarius, Capricorn, Aquarius, and Pisces. The names of those constellations are what astrology buffs call the signs of the Zodiac. (Actually, a 13th constellation, Ophiuchus, intersects the ecliptic, but in ancient times, it wasn't included in the Zodiac, so it's not a sign.)

Earth's steady progression along its orbit of the Sun results in a different appearance of the night sky over the course of the year. (The same effect is what makes the Sun move around the circle of the ecliptic; as Earth moves, we see the Sun in different directions against the background of fixed stars.) The stars aren't located in the same places with respect to the horizon throughout the night or throughout a year (the North Star is an exception; it is in nearly the same place all night and every night). The constellations that appeared high in the sky at dusk a month ago are lower in the west after sunset now. And if you view the constellations that loom low in the east just before dawn, you preview what you'll see at midnight in a few months.

To keep track of the constellations, use the sky maps that come monthly in astronomy magazines such as *Sky & Telescope* and *Astronomy*. (For more about magazines, check out Chapter 2.) Your daily newspaper may also contain maps of the evening skies, or you can get an inexpensive *planisphere,* or star wheel, which features a rotating disk in a square frame, with a hole cut into it that represents the limits of your view. You want a planisphere designed for your latitude, or roughly so.

Here are two good planispheres to consider:

- *The Night Sky,* by David Chandler, available in five versions, for various latitudes, and in two different sizes. Get the version that corresponds to your latitude, and pick the large size.

- *David H. Levy's Guide to the Stars,* bigger than most other planispheres, is suitable for most places in the United States, the United Kingdom, and elsewhere between latitudes 30° and 60° north.

Planispheres are available from The Sky & Telescope Shop (`www.shopat sky.com`), Astronomics (`www.astronomics.com`), and the Astronomical League's online store (`www.astroleague.org/store/index.php`). You can also find them at Amazon.com and similar vendors.

You can even order a planisphere watch from Edmund Scientifics (`www. scientificsonline.com`). (Wait until somebody stops you on the street and asks what time you have on your watch, and you reply, "Well, it's half past Leo the Lion!")

Using a planisphere helps you understand the motions of the stars as you turn the disk to show the stars for different hours and dates. But if you don't care about learning that stuff and just want to know what constellations are up in the sky, you're better off downloading a cheap or free planisphere app (also called a planetarium app) for your smartphone or tablet computer, as I describe in Chapter 2.

. . . keep an eye on the North Star

Anyone can walk outside on a clear night and see some stars. But how do you know what you're seeing? How can you find it again? What do you watch for?

One of the most time-honored ways of getting familiar with the night sky if you live in the Northern Hemisphere is to pay attention to the North Star, or Polaris, which barely moves. After you identify which way is north, you can orient yourself to the rest of the northern sky — or to anywhere else, for that matter. In the southern sky, you need to find the bright stars Alpha and Beta Centauri (a Southern Hemisphere planisphere, smartphone app, or simple star map can help), which point the way to the Southern Cross.

You can easily find the North Star by using the Big Dipper in the constellation Ursa Major (see Figure 3-1). The Big Dipper is one of the most easily recognized sky patterns. If you live in the continental United States, Canada, or the United Kingdom, you can see it every night of the year.

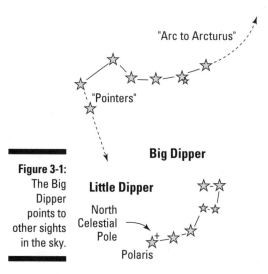

"Arc to Arcturus"

"Pointers"

Big Dipper

Figure 3-1:
The Big Dipper points to other sights in the sky.

Little Dipper

North Celestial Pole

Polaris

The two brightest stars in the Big Dipper, Dubhe and Merak, form one end of its bowl and point directly to the North Star. The Big Dipper helps you locate the bright star Arcturus, in Bootes, too: Just imagine a smooth continuation of the curve of the Dipper's handle, as in Figure 3-1.

The stars close to Polaris never set below the horizon in the United Kingdom, nor at most latitudes in North America, which makes them *circumpolar stars:* They appear to circle around Polaris. Ursa Major is a circumpolar constellation seen from almost all the Northern Hemisphere. The circumpolar area of the sky depends on your latitude. The closer you live to the North Pole, the more of the sky is circumpolar. And in the Southern Hemisphere, the farther south your location, the greater the circumpolar part of the sky. But if a constellation is circumpolar in the Northern Hemisphere, it cannot be circumpolar in the Southern Hemisphere, and vice versa.

Orion is a distinctive constellation visible in the Northern Hemisphere's winter evening sky, with the three stars that make up its belt pointing in one direction to Sirius in Canis Major and in the opposite direction to Aldebaran in Taurus. Orion also contains the 1st magnitude stars Betelgeuse and Rigel, two brilliant beacons in the sky (see Figure 3-2). For more on the magnitude of stars, head to Chapter 1.

Figure 3-2:
Orion and
its bright,
beacon-
like stars
Rigel and
Betelgeuse.

Betelgeuse

Orion

Rigel

You can become friends with the night sky by looking at this book's constellation maps (check out Appendix A) and checking them out with your own eyes. Just as becoming familiar with the streets of your city helps you find your way faster, knowing the constellations helps you set your sights on the sky objects you want to observe. Gaining sky knowledge also assists you in tracking the appearance of the stars and their movements as you maneuver through a nightly session.

Beginning with Naked-Eye Observation

If you don't already know the compass directions in your area, take the time to familiarize yourself with them. You need to know north from south and east from west. When you get your bearings, you can use the weekly sky highlights from the *Sky & Telescope* website (www.skyandtelescope.com) or use a planisphere, a planetarium-type smartphone app, or a desktop planetarium program to orient yourself to the brightest stars and planets in the sky at the time you look. (For more about these astronomy resources, see Chapter 2 and also flip earlier in this chapter.) When you recognize the bright stars, you have an easier time picking out the patterns of slightly fainter ones all around them.

Table 3-1 lists some of the brightest stars you can see in the night sky, the constellations that contain them, and their magnitudes (the measure of their brightness — see Chapter 1). Many of them are visible from the continental United States, the United Kingdom, and Canada. Some you can see well from only southern latitudes, so bright stars that you can't see in the United States may be prominent sights for Australians. See Chapter 11 for information on spectral class, which is an indication of the color and temperature of a star. (Spectral class B stars, for example, are white and rather hot, and M stars are red and relatively cool.)

Start your observations by consulting one of the sky map resources I describe earlier and see how many of the brightest stars you can locate at night. After you do, try identifying some of the dimmer stars in the same constellations. And, of course, keep your eye out for the bright planets: Mercury, Venus, Mars, Jupiter, and Saturn (which I cover in Chapters 6 and 8).

Table 3-1	The Brightest Stars As Seen from Earth		
Common Name	**Apparent Magnitude**	**Constellation Designation**	**Spectral Class**
Sirius	−1.5	α Canis Majoris	A
Canopus	−0.7	α Carinae	A
Rigil Kentaurus	−0.3	α Centauri	G
Arcturus	−0.04	α Bootis	K
Vega	0.03	α Lyrae	A
Capella	0.1	α Aurigae	G
Rigel	0.1	β Orionis	B
Procyon	0.4	α Canis Minoris	F
Achernar	0.5	α Eridani	B
Betelgeuse	0.5	α Orionis	M
Hadar	0.6	β Centauri	B
Acrux	0.8	α Crucis	B
Altair	0.8	α Aquilae	A
Aldebaran	0.9	α Tauri	K
Antares	1.0	α Scorpii	M
Spica	1.0	α Virginis	B
Pollux	1.1	β Geminorum	K
Fomalhaut	1.2	α Piscis Austrini	A
Deneb	1.3	α Cygni	A

In winter and summer, the Milky Way runs high in the sky at most locations in the Northern Hemisphere. If you can recognize the Milky Way as a wide, faintly luminous band across the sky, you have at least a pretty fair observing site. And if the Milky Way seems bright, you must be at a dark sky location like those that I describe in Chapter 2.

The most important step in observing with the naked eye is to shield your vision from interfering lights. If you can't get to a dark place in the country, find a dark spot in your backyard or possibly on the roof of a building. You won't eliminate the light pollution high up in the sky that results from the collective lights of your city, but trees or the wall of a house can prevent nearby lights, including streetlamps, from shining in your eyes. After 10 or 20 minutes, you can see fainter stars; you're getting "dark adapted."

Watching the beautiful comet Hyakutake in 1996 from a small city in the Finger Lakes of northern New York, I found that walking around the corner of a building to shield myself from nearby street lights made the comet a much better sight.

Ideally, you want a site with a good horizon and only trees and low buildings in the distance, but finding that kind of location is next to impossible in a major urban area.

If you fail to find a site with a good horizon in all directions, the most important horizon is the southern one (if you live in the Northern Hemisphere). You make most observations in the Northern Hemisphere while facing roughly south (with east to the left and west to the right). As you face south, the stars rise to your left and set to your right. If you observe from the Southern Hemisphere, reverse this procedure and face north: The stars rise in the east (on your right) and set in the west (on your left).

How bright is bright?

I talk about magnitude in Chapter 1, but it helps to know that astronomers can define magnitude in different ways for different purposes:

✔ **Absolute magnitude** is the brightness of a sky object as seen from a standard distance of 32.6 light-years. Astronomers consider it the "true" magnitude of the object.

✔ **Apparent magnitude** is how bright an object appears from Earth, which is usually different than its absolute magnitude, depending on how far away from Earth the sky object is located. A star closer to Earth may appear brighter than one farther away, even if its absolute magnitude is fainter.

✔ **Limiting magnitude** is the apparent magnitude of the faintest star that you can see. It depends on how clear the sky is at the time of observation and how dark the sky is. A very bright star may be invisible if many clouds hang overhead, for example, and city lights or a full Moon may interfere with viewing fainter stars that you can see with the naked eye under good conditions. Limiting magnitude is especially important in meteor and deep sky observations. On a clear, dark night, the limiting magnitude may be 6 at the zenith, but in the city, the limiting magnitude may be only 3 or 4.

Star charts depict the apparent magnitudes of the stars to simulate their appearance in the sky.

Always have a watch, a notebook, and a dim or red flashlight to use for recording what you see. Some flashlights come with a red bulb, or you can buy red cellophane from a greeting card store to wrap around the lamp. After you become dark adapted, white light reduces your ability to see the faint stars, but dim red light doesn't hurt your dark adaptation.

Using Binoculars or a Telescope for a Better View

As with any new hobby, you want to gain some experience and research what's available before you start buying expensive equipment. You don't want to buy a telescope until you've seen several telescopes of different types in action and discussed them with other observers. In the following sections, I offer my advice for choosing the right binoculars or telescope for you.

Don't even think about looking at the Sun with a telescope or binoculars unless you use the safe procedures and special equipment that I mention in "Staying safe when you view the Sun," later in this chapter, and that I describe in detail in Chapter 10. Otherwise, you can suffer permanent harm.

Binoculars: Sweeping the night sky

Owning a good pair of binoculars is a must. Buy or borrow a pair and observe with them before you get a telescope. Binoculars are excellent for many kinds of observation, and if you give up on astronomy (sigh), you can still use them for many other purposes. And if you borrowed the binoculars, be sure to return them before you are viewed with suspicion.

Binoculars are great for observing variable stars, searching for bright comets and novae, and sweeping the sky just to enjoy the view. You may never discover a comet yourself, but you'll certainly want to view some of the brighter ones as they appear. Nothing works better for this purpose than a good pair of binoculars.

The following sections cover the way binoculars are specified according to their capabilities, and I show you the steps to take as you figure out what kind of binoculars to buy. Figure 3-3 takes you inside a pair of binoculars.

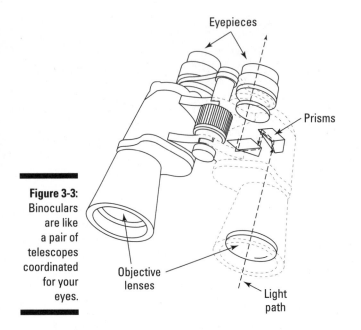

Eyepieces

Prisms

Figure 3-3:
Binoculars
are like
a pair of
telescopes
coordinated
for your
eyes.

Objective
lenses

Light
path

Prisms, glass, and shapes

Binoculars contain prisms to bend the light coming from the two large lenses *(objective lenses)* to the two smaller lenses *(eyepieces)* that you look through. It's necessary because the eyepieces can't be farther apart than the distances between your eyes, or you won't be able to look through both eyepieces at once. The objectives are bigger than your eyes, so they need to be farther apart; therefore, the light paths through the binoculars have to be bent.

The two basic prism types in binoculars are as follows:

- ✓ **Roof prisms** are used in binoculars that are relatively straight and narrow; these binoculars are favorites among bird-watchers.

- ✓ **Porro prisms** are used in binoculars that are relatively wide and short; they are the better type for stargazing because they give brighter images for the same-size lenses. It's also easier to hold wide binoculars steady.

Binoculars use two main types of glass:

- ✓ **BK-7 glass,** a trade term for garden-variety borosilicate glass, is often used in cheap binoculars.

- ✓ **BaK-4 glass,** or barium crown glass, is used in fine binoculars and often yields brighter images of dim astronomical objects.

Deciphering the numbers on binoculars

Binoculars come in many sizes and types, but each pair of binoculars is described by a numerical rating — 7×35, 7×50, 16×50, 11×80, and so on. (Note that the ratings read "7 by 35, 7 by 50, 16 by 50, and 11 by 80." Don't say "7 times 35.") Here's how to decode these ratings:

- ✔ The first number is the optical magnification. A 7×35 or 7×50 pair of binoculars makes objects look seven times larger than they do to the naked eye.

- ✔ The second number is the *aperture,* or diameter, of the light-collecting lenses (the big lenses) in the binoculars, measured in millimeters. An inch is about 25.4 millimeters; thus, 7×35 and 7×50 binoculars have the same magnifying power, but the 7×50 pair has bigger lenses that collect more light and show you fainter stars than the 7×35 pair.

Also keep the following considerations in mind:

- ✔ Bigger binoculars reveal fainter objects than smaller ones do, but they weigh more and are harder to hold and point steadily toward the sky.

- ✔ Higher-magnification binoculars, such as 10×50 and 16×50, show objects with greater clarity, provided that you can hold them steady enough, but they have smaller fields of view, so finding celestial targets is harder than with lower-magnification binoculars.

- ✔ Giant binoculars — 11×80, 20 ×80, and on up — are heavy and hard to hold steady; many people can't use them without a tripod or stand. The very biggest, 40×150, must be used with a stand.

- ✔ Many intermediate sizes are available, such as 8×40 or 9×56.

Here's my opinion: 7×50 is the best size for most astronomical purposes and certainly the best size to start with. If you purchase binoculars much smaller than 7×50, you really equip yourself for bird-watching rather than astronomy. Most astronomers can use 7×50 binoculars without a tripod or stand, although some people may need to brace themselves to hold these binoculars steady.

Buy a much larger size than 7×50, and you may be investing in a white elephant that you'll rarely use.

Making sure your binoculars are right for you

First and foremost, don't buy binoculars unless you can return them after a trial run. Here's how to make the basic check to determine whether a pair of binoculars is worth keeping:

✔ The image should be sharp across the field of view when you look at a field of stars.

✔ You should have no difficulty focusing the binoculars for your eyesight, with a separate adjustment for at least one of the eyepieces (the small lenses next to your eyes when you look through the binoculars).

✔ When you adjust the focus, it should change smoothly. Stars' images should be sharp points when in focus and circular in shape when not.

✔ Special transparent coatings are deposited on the objective lenses (large lenses) of many binoculars. This feature, called *multicoating,* results in a clearer, more contrasting view of star fields. Binoculars that are *fully multicoated* are even better; in them, the coatings are applied to all the lenses and prisms.

Some astronomers wear their eyeglasses when observing with binoculars. Others, like me, are more comfortable putting away the glasses when using binoculars. But if you don't have your eyeglasses on, you may have a problem writing notes, reading a star chart, and so on. Your choice of the best binoculars for you may depend on whether you will keep your glasses on, as I explain next.

If you plan to wear your eyeglasses when you observe with binoculars, you need binoculars that have enough *eye relief.* Eye relief is the distance (measured in millimeters) from the outer surface of a binocular eyepiece to the *focal point,* where the binoculars focus an image. If your eye is beyond that distance from the eyepiece, you can't see the whole field of view. That circumstance happens when the thickness of your eyeglasses keeps your eye from getting to the focal point. Here's my advice: Disregard salesperson assurances and binoculars manufacturers' specs on eye relief. Instead, perform this simple test on binoculars that you are considering buying:

1. **Take off your glasses, and focus the binoculars on a scene a block away (or in the sky).**

   Note how much of the scene is in view.

2. **Put on your glasses.**

   If the eyepieces have rubber eye cups, fold down the cups so that you can get closer to the eyepieces while wearing glasses.

3. **Focus the binoculars on the same scene as before.**

   If you don't see as much of the scene with your glasses on, the binoculars don't have enough eye relief.

Good binoculars are sold in optical and scientific specialty stores. Some large camera stores have decent binocular selections. But I recommend avoiding department stores. You may get low-grade merchandise in some department stores or pay exorbitant prices for fancy binoculars in others. And you can bet that the salespeople peddling them know less than you do.

You can pay hundreds of dollars or even a few thousand bucks for a good pair of 7×50 binoculars, but if you shop around, you can find a perfectly adequate pair for $120 or less. (Pawn shops and military surplus stores are excellent places to look.) Used binoculars are often a good deal, but you must try them before you buy them because they may be out of whack.

Many astronomers buy their binoculars from specialty retailers and manufacturers that advertise in astronomy magazines and on the web (see Chapter 2 for more about these resources). If you must order by mail (or from the web), ask experienced amateurs whom you meet at an astronomy club or ask a staff member at a planetarium to recommend a dealer.

Reputable makers of binoculars include Bushnell, Canon, Celestron, Fujinon, Meade, Nikon, Orion, Pentax, and Vixen. Some high-end Canon and Nikon binoculars have image stabilization, a high-tech feature that makes the image much steadier. They come in handy on a boat rocking at sea and are often a great help on land as well.

Telescopes: When closeness counts

If you want to look at the craters on the Moon, the rings of Saturn, or Jupiter's Great Red Spot (all of which I describe in Part II), you need a telescope. The same advice goes for observing faint variable stars or viewing all but the brightest galaxies or the beautiful small glowing clouds called *planetary nebulae,* which have nothing to do with planets (see Chapters 11 and 12).

Before you view the Sun or any object crossing in front of the Sun, however, be sure to read the special instructions in Chapter 10 to protect your eyes and avoid going blind!

The following sections cover telescope classifications, mounts, and shopping tips to find the best telescope for your needs.

Focusing on telescope classifications

Telescopes come in three main classifications:

- Refractors use *lenses* to collect and focus light (see Figure 3-4). In most cases, you look straight through a refractor.
- Reflectors use *mirrors* to collect and focus light (see Figure 3-5). Reflectors come in different types:
  - In a *Newtonian* reflector, you look through an eyepiece at right angles to the telescope tube.
  - In a *Cassegrain* telescope, you look through an eyepiece at the bottom.

- A *Dobsonian* reflector gives you the most *aperture* (or light-gathering power) for your money, but you may have to stand on a stool or a ladder to look through it. Dobsonians tend to be larger than other amateur telescopes (because large Dobsonians are more affordable), and the eyepiece is up near the top.

✔ *Schmidt-Cassegrains* and *Maksutov-Cassegrains* use both mirrors and lenses. These models are more expensive than reflectors with comparable apertures.

Many varieties are available within these general telescope types. And every telescope used for amateur purposes is equipped with an *eyepiece,* which is a special lens (actually, a combination of lenses mounted together as a unit) that magnifies the focused image for viewing. When you take photographs, you usually remove the eyepiece and mount the camera on the telescope in its place.

Just as with a microscope or a camera with interchangeable lenses, you can use interchangeable eyepieces with almost any telescope. Some companies don't make telescopes at all, specializing instead in making eyepieces that work with many different telescopes.

Beginners usually buy the highest-magnification eyepieces they can, which is a great way to waste your money. I recommend low- and medium-power eyepieces because the higher the power, the smaller the field of view, making it tougher to track faint (and possibly even bright) targets. For a small telescope, observation is usually best with eyepieces that give magnifications of 25x or 50x, not 200x or more. (The *x* stands for "times," as in 25 times bigger than what you see with the naked eye.) If you see a telescope advertised for its "high power," the advertisers may be trying to sell mediocre goods to unwitting buyers. And if a salesperson touts the high power of a telescope, patronize another store.

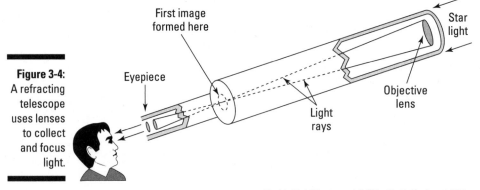

Figure 3-4:
A refracting telescope uses lenses to collect and focus light.

First image formed here

Star light

Eyepiece

Objective lens

Light rays

Dinah L. Moché/Astronomy: A Self-Teaching Guide, Seventh Edition

Calculating the magnification of your eyepiece

Eyepieces are usually marked with their focal lengths in millimeters, and a given eyepiece may provide different magnifications with different telescopes. If you know the focal length of your telescope, you can calculate the magnification that a particular eyepiece provides. Follow these steps to determine the magnification your eyepiece provides for your telescope:

1. **Determine the focal length of the telescope.**

   If your telescope lists the focal length, make note of it and skip to Step 3.

   To calculate the focal length, multiply the f/number listed on the telescope, such as f/5.6 or f/8, by the diameter of the main lens or mirror. For example, if the diameter is 6 inches and the f/number is f/8, the focal length is 6 × 8 = 48 inches.

2. **Convert the focal length from inches to millimeters.**

   Multiply the length in inches by 25.4 (the number of millimeters per inch). For example, if your focal length is 48 inches, 48 × 25.4 = 1,219.2 millimeters.

3. **Divide the focal length of your telescope by the focal length of the eyepiece.**

   If the telescope focal length is 1,219.2 millimeters and the eyepiece focal length is 25 millimeters, the eyepiece provides a magnification of approximately 49 times (1,219.2 ÷ 25 = 48.768).

Figure 3-5:
English scientist Sir Isaac Newton invented the reflector, which uses mirrors to collect and focus light.

Eyepiece

Star light

Primary mirror

Prime focus

Dinah L. Moché/Astronomy: A Self-Teaching Guide, Seventh Edition

What limits your view of fine details with a small telescope isn't the power of the eyepiece. Air turbulence (the same factor that makes stars seem to twinkle) and any shaking of the telescope in the breeze are the effects that limit the clarity of your view.

Coloring your universe

What do you see when you look at a sky object through your binoculars or telescope? Do you see glorious stars, planets, and sky objects in brilliant colors, as shown in the photos in the color section of this book? Not!

Sorry for the tease, but you're likely to see most stars and sky objects in pale colors. Most stars look white or off-white by eye, binoculars, or telescope — yellowish rather than yellow, for example. The colors are most vivid when adjacent stars have contrasting colors, as found in telescopic views of some double stars.

Most photos of sky objects have been color enhanced, traditionally described as having *false color*. Astronomers don't use false color to gussy up the universe, which is fine as it is.

Nor is it meant to give a false impression of the deep sky. In fact, the enhancement furthers the search for truth, much like a stain on medical slides brings out the detail in the cells and helps identify physical differences and relationships.

Depending on the method of observation and presentation, photos of the same object can be strikingly different. But they all tell scientists about differences in the structure of the object, what substances it may contain, and what dynamic processes are taking place. Also, many astronomical images are obtained in forms of light invisible to the human eye (like ultraviolet, infrared, and X-ray), so astronomers use false color in this absence of any recognizable color.

Examining telescope mounts

Telescopes are generally mounted on a stand, a tripod, or a pier in one of two ways:

- ✔ With an *alt-azimuth mount,* you can swivel the telescope up and down and side to side — in altitude (the vertical plane) and azimuth (the horizontal plane). You need to adjust the telescope on both axes to compensate for the motion of the sky as Earth rotates. Dobsonian reflectors always use alt-azimuth mounts.

- ✔ With the more expensive *equatorial mount,* you align one axis of the telescope to point directly at the Celestial North Pole or, for Southern Hemisphere viewers, the Celestial South Pole. After you spot an object, simply turning the telescope around the polar axis keeps it in view. Be sure to polar-align the scope for each viewing session.

An alt-azimuth mount is usually steadier and easier for a beginner to use, but an equatorial mount is better for tracking the stars as they rise and set. However, if a telescope has built-in computer control (as in both models that I recommend shortly), either style of mount is fine because the computer takes care of tracking.

The objects you see in the telescope are usually upside down, which isn't the case with binoculars. Of course, it doesn't really make much difference in the viewing, but just know that top and bottom are reversed when you view through a telescope. Adding a lens to rotate the image so that it's right side up has the disadvantage that it reduces the amount of light coming through the scope and dims the image, so it's best not done. When viewed through an equatorially mounted telescope, a star field maintains the same orientation as it rises and sets. But when seen through an alt-azimuth mount telescope, the field rotates during the night, so the stars on top wind up on the side.

Shopping for telescopes the smart (and economical) way

A cheap, mass-manufactured telescope, often called a *drugstore* or *department store* telescope, is usually a waste of money. And it still costs more than a hundred or maybe several hundred dollars.

A good telescope, bought new, may run you several hundred dollars to $1,000, and you certainly can pay more. But you can find alternatives:

- ✔ Used telescopes are often sold through ads in astronomy magazines or in the newsletters of local astronomy clubs. If you can inspect and test a used telescope and you find what you like, buy it! A well-maintained telescope can last for decades.

- ✔ In many areas, amateurs can observe with the larger telescopes operated by astronomy clubs, planetariums, or public observatories.

The technology of amateur telescopes is advancing at a rapid pace, and a former astronomer's dream can be today's obsolete equipment. Quality and capabilities are going up, and prices are generally fair, perhaps because reliable manufacturers are in competition for your money.

Generally, a good refractor gives better views than a good reflector that has the same *aperture,* or telescope size. Aperture refers to the diameter of the main lens, the mirror, or, in a more complicated telescope, the size of the unobstructed portion of the optics. But a good refractor is more expensive than a comparable reflector.

Staying safe when you view the Sun

Taking even the briefest peek at the Sun through a telescope, binoculars, or any other optical instrument is dangerous unless the device is equipped with a solar filter made by a reputable manufacturer specifically for viewing the Sun. And the filter must be properly mounted on the telescope, not jury-rigged.

You must also use a solar filter when you view Mercury as it crosses *(transits)* the disk of the Sun. (I describe transits of Mercury in Chapter 6.) Viewing an object against the Sun requires the use of protective viewing techniques because you're also looking at the Sun. If you have a Newtonian reflector, a Dobsonian reflector, or a refractor, you can try using projection. See Chapter 10 for specifics about solar viewing and how to protect your eyes.

Good seeing gone bad

Turbulence in the atmosphere affects how well you can see the stars. It makes the stars seem to twinkle. The term *seeing* describes the conditions of the atmosphere relating to steadiness of the image — *good seeing* is when the air is stable and the image holds steady. You often have better seeing late at night, when the heat of the day has dissipated. When seeing is bad, the image tends to "break up," and double stars blur together in telescopic views. The stars always twinkle most close to the horizon, where the seeing is worst.

The bright planets Mercury, Venus, Mars, Jupiter, and Saturn generally don't twinkle when you view them with the naked eye. They are not single points of light like a star seen from Earth, but are seen as disks. Each disk consists of many individual points of light. The separate points twinkle but average each other out (one twinkling brighter, another twinkling fainter at any moment), so you see a steady light from the planet.

The warmth of a telescope brought out from a heated home into the cool night air causes some bad seeing; wait a while for the telescope to cool down, and your viewing will improve. Situations vary, but 30 minutes is usually enough to make a significant difference in the viewing.

The Maksutov-Cassegrains and Schmidt-Cassegrains are good compromises between the low cost of a reflector and the high performance but high cost of a refractor. For many astronomers, these hyphenated telescope types are the preferred models.

One of the best small telescopes is the Meade ETX-90. Its aperture is 3.5 inches, almost the smallest size of any telescope that you should start with. (If you find a good instrument at a good price from a 2.5-inch aperture on up, especially in a refractor, consider it for purchase.)

The ETX-90 sells for about $400 and comes with an Autostar computerized controller and a tripod. This instrument automatically points at almost any object you specify if that object is in view from your location at that time. The Autostar can even find moving objects, such as planets, based on stored information, and it's equipped to give you a "tour" of the best sights in the sky, selected with no input from you.

A good competing telescope for the ETX-90 is the Celestron SkyProdigy 90. It's comparably sized and equipped and has the capability to automatically align itself on the sky, after which it points to almost any celestial object that you select. It goes for about $600.

You definitely don't want to spend this much money on a telescope until you see the same model in action at an astronomy club observing meeting or a star party (see Chapter 2). But the price is no more than you pay for a fine camera and an accessory lens or two. You can find larger telescopes for less money — check the ads in current issues of astronomy magazines — but you have to invest much more effort in learning to use them effectively.

Some brand-name telescopes are sold through authorized dealers that tend to have expert knowledge. But take their advice with just a wee bit of salt.

Here are some key websites to browse for telescope product information:

- ✔ Celestron, for many years the favorite manufacturer for thousands of astronomers (www.celestron.com)
- ✔ Meade Instruments Corporation (www.meade.com)
- ✔ Orion Telescopes & Binoculars (www.telescope.com)

On each of these websites, you can find the instruction manuals for many of the telescopes that they sell. Consider taking a look at the manual before you buy a telescope so that you know whether it will be helpful when you run into problems.

If you don't live in the United States, you may be able to find dealers for the telescope brands in your country, The Widescreen Centre, in London, England, carries Celestron, Meade, and Orion telescopes (www.widescreen centre.co.uk). In Australia, The Binocular and Telescope Shop, with locations in Sydney and Melbourne, is among the dealers that carry these telescopes (www.bintel.com.au).

Planning Your First Steps into Astronomy

I recommend that you get into the astronomy hobby gradually, investing as little money as possible until you're sure about what you want to do. Here's a plan for acquiring both basic skills and the needed equipment:

1. **If you have a late-model computer, invest in a free or inexpensive planetarium program. Better yet, if you have a smartphone, download and use a free or cheap planetarium app (see recommendations for apps and programs in Chapter 2). Start making naked-eye observations at dusk on clear nights and before dawn, if you're an early riser.**

   To plan your observations of planets and constellations, you can also rely on the weekly sky scenes at the *Sky & Telescope* website (www.skyandtelescope.com). If you don't have a suitable computer, plan your observations based on the monthly sky highlights in *Astronomy* or *Sky & Telescope* magazines.

2. **After a month or two of familiarizing yourself with the sky and discovering how much you enjoy it, invest in a serviceable pair of 7×50 binoculars.**

3. **As you continue to observe the bright stars and constellations, invest in a star atlas that shows many of the dimmer stars, as well as star clusters and nebulae.**

Sky & Telescope's Pocket Star Atlas, by Roger W. Sinnott (Sky & Telescope Media, 2006) is a good choice. Compare scenes in your star atlas with the constellations that you're observing; the atlas shows their RAs and Decs (see Chapter 1 for info about RAs and Decs). Eventually, you'll start to develop a good feel for the coordinate system.

4. **Join an astronomy club in your area, if at all possible, and get to know the folks who have experience with telescopes. (See Chapter 2 for more about finding clubs.)**

5. **If all goes well and you want to continue in astronomy — as I bet you will — invest in a well-made, high-quality telescope in the 2.5- to 4-inch size range.**

   Study the telescope manufacturer websites earlier in this chapter or send for catalogs advertised in astronomy magazines. Better yet, talk to experienced astronomy club members, if you can. They can advise you on buying a new telescope, and they may know someone who wants to sell a used telescope.

If you find that you enjoy astronomy as much as I think you will, after a few years, consider moving up to a 6- or 8-inch telescope. It may be harder to use, but you'll be ready to master it after you have some experience. Equipped with a larger telescope, you can see many more stars and other objects. You can get ideas about what larger telescopes to consider by talking to other amateur astronomers and by attending a star party, where you can see many different telescopes in operation and on display. (I cover star parties in Chapter 2.)

Chapter 4

Just Passing Through: Meteors, Comets, and Artificial Satellites

See a moving object in the daytime sky? You probably know whether it's a bird, a plane, or Superman. But in the night sky, can you distinguish a meteor from a flash of light glinting off an Iridium satellite? And among objects that move slowly but perceptibly across the starry background, can you tell a comet from an asteroid?

This chapter defines and explains many of the objects that sweep across the night sky. (The Sun, Moon, and planets move across the sky, too, but in a more stately procession. I focus on them in Parts II and III.) When you know how to identify these night visitors, you can look forward to enjoying them all.

Meteors: Wishing on a Shooting Star

No astronomy term is misused more often than the word *meteor*. Amateur astronomers and even scientists are quick to spurt out *meteor* when *meteoroid* or *meteorite* is the accurate term. Take a look at the correct meanings:

✔ A *meteor* is the flash of light produced when a naturally occurring small, solid object (a meteoroid) enters Earth's atmosphere from space; people often call meteors "shooting stars" or "falling stars."

✔ A *meteoroid* is a small, solid object in space, usually a fragment from an asteroid or comet, that orbits the Sun. Some rare meteoroids are actually rocks blasted off Mars and Earth's Moon.

✔ A *meteorite* is a solid object from space that has fallen to the surface of Earth. About 100 tons of meteoritic material fall on Earth every day (some estimates are even higher).

If a meteoroid runs into Earth's atmosphere, it may produce a meteor bright enough for you to see. If the meteoroid is big enough to hit the ground instead of disintegrating in midair, it becomes a meteorite. Many people hunt for and collect meteorites because of their value to scientists and collectors.

The two main kinds of meteoroids have different places of origin:

✔ *Cometary meteoroids* are fluffy little dust particles shed by comets.

✔ *Asteroidal meteoroids,* which range in size from microscopic particles to boulders, are literally chips from asteroids — the so-called minor planets — which are rocky bodies that orbit the Sun (and which I describe in Chapter 7).

When you go to a science museum and see a meteorite on display, you're examining an asteroidal meteoroid that fell to Earth (or, in rare cases, a rock that fell after being knocked off the Moon or Mars by a larger impacting body). It may be made of stone, iron (actually, an almost rustproof mixture of nickel and iron), or both. Showing rare simplicity (for once), scientists call these meteorite types, respectively, *stony, iron,* and *stony-iron meteorites.*

Comb yourself for space dust

If an astronomer finds a *micrometeorite* (a meteorite so small that you must view it through a microscope), it may be a particle that began as a cometary meteoroid, or it may be a very small asteroidal meteoroid.

Micrometeorites are so small that they don't create enough friction to burn up or disintegrate in the atmosphere, so they sift slowly down to the ground. Chances are, you have one or two pieces of this space dust in your hair right now, but the dust is almost impossible to identify because it would be lost among the millions of other microscopic particles on your head (no offense).

Scientists obtain micrometeorites by flying ultraclean collecting plates on high-altitude jet aircraft. And they drag magnetized rakes, which pick up micrometeorites made of iron, through the mud on the sea floor.

On January 2, 2004, NASA's Stardust space probe flew past Comet Wild-2 (a comet that passes inside the orbit of Mars once every six years or so and is fairly easy to reach with a probe) and collected some comet dust. The probe ended its "wild ride" by ejecting a capsule containing the dust it had collected. It parachuted down in Utah on January 15, 2006. A team of 200 scientists analyzed the tiny particles. They found that some dust grains came from other stars, but most formed near the Sun, some so close that, according to Stardust scientist Donald Brownlee, it was "hot enough to evaporate bricks." The probe, renamed Stardust-NeXT, flew on to photograph Comet Tempel 1 on Valentine's Day in 2011. You can see many of the photos at http://star dustnext.jpl.nasa.gov/multi media/tempel1_images.html.

In the following sections, I cover three types of meteors: sporadic meteors, fireballs, and bolides. I also give you the scoop on meteor showers.

For reliable information on how to observe, record, and report meteors, visit the "Meteor" section of the British Astronomical Association website, at www. britastro.org/meteor. The North American Meteor Network is another good resource. It provides a beginner's guide to meteor observing, forms to submit your meteor counts, and special forms for reporting fireballs. Check out the website at www.namnmeteors.org. While you're surfing the web, head over to the International Meteor Organization's website, at www.imo. net, to see what it has to offer.

Spotting sporadic meteors, fireballs, and bolides

When you're outdoors on a dark night and see a "shooting star" (the flash of light from a random, falling meteoroid), what you're probably seeing is a *sporadic* meteor. But if many meteors appear, all seeming to come from the same place among the stars, you're witnessing a *meteor shower.* Meteor showers are among the most enjoyable sights in the heavens; I devote the next section in this chapter to them.

A dazzlingly bright meteor is a *fireball.* Although a fireball has no official definition, many astronomers consider a meteor that looks brighter than Venus to be a fireball. However, Venus may not be visible at the time you see the bright meteor. So how can you decide whether you're seeing a fireball?

Here's my rule for identifying fireballs: If people facing the meteor all say "Ooh" and "Ah" (everyone tends to shout when they see a bright meteor), the meteor may be just a bright one. But if people who are *facing the wrong way* see a momentary bright glow in the sky or on the ground around them, it's the real thing. To paraphrase an old Dean Martin tune, when the meteor hits your eye like a big pizza pie, that's a fireball!

Fireballs aren't very rare. If you watch the sky regularly on dark nights for a few hours at a time, you'll probably see a fireball about twice a year. But *daylight fireballs* are very rare. If the Sun is up and you see a fireball, mark it down as a lucky sighting. You've seen one tremendously bright fireball. When nonscientists see daytime fireballs, they almost always mistake them for an airplane or missile on fire and about to crash.

Any very bright fireball (approaching the brightness of the half Moon or brighter) or any daylight fireball represents a possibility that the meteoroid producing the light will make it to the ground. Freshly fallen meteorites are often of considerable scientific value, and they may be worth good money, too. If you see a fireball that fits this description, write down all the following information so that your account can help scientists find the meteorite and determine where it came from:

1. **Note the time, according to your watch.**

   At the earliest opportunity, check how fast or slow your watch is running against an accurate time source, such as the Master Clock at the U.S. Naval Observatory, which you can consult at `http://tycho.usno.navy.mil/what.html`.

2. **Record exactly where you are.**

   If you have a Global Positioning System receiver handy, take a reading of your location. Otherwise, make a simple sketch showing where you stood when you saw the fireball — note roads, buildings, big trees, or any other landmarks.

3. **Make a sketch of the sky, showing the track of the fireball with respect to the horizon as you saw it.**

   Even if you're not sure whether you faced southeast or north-northwest, a sketch of your location and the fireball track helps scientists determine the trajectory of the fireball and where the meteoroid may have landed.

After a daylight fireball or a very bright nighttime fireball, interested scientists advertise for eyewitnesses. They collect the information, and by comparing the accounts of persons who viewed the fireball from different locations, they can close in on the area where it most likely fell to the ground. Even a brilliant fireball may be only the size of a small stone — one that would fit easily in the palm of your hand — so scientists need to narrow the search area to have a reasonable chance of finding it. If you don't see a call for information after your fireball observation, chances are good that the nearest planetarium or natural history museum will accept your report and know where to send it.

A *bolide* is a fireball that explodes or that produces a loud noise even if it doesn't break apart. At least, that's how I define it. Some people use *bolide* interchangeably with *fireball.* (You won't find an official agreement on this term; you can find different definitions in even the most authoritative sources.) The noise that you hear is the sonic boom from the meteoroid, which is falling through the air faster than the speed of sound.

When a fireball breaks apart, you see two or more bright meteors at once, very close to each other and heading the same way. The meteoroid that produces the fireball has fragmented, probably from aerodynamic forces, just as an airplane falling out of control from high altitude sometimes breaks apart even though it hasn't exploded.

Often a bright meteor leaves behind a luminous track. The meteor lasts a few seconds or less, but the shining track — or *meteor train* — may persist for many seconds or even minutes. If it lasts long enough, it becomes distorted by the high-altitude winds, just as the wind gradually deforms the skywriting from an airplane above a beach or stadium.

You see more meteors after midnight local time than before because, from midnight to noon, you're on the forward side of Earth, where our planet's plunge through space sweeps up meteoroids. From noon to midnight, you're on the backside, and meteoroids have to catch up in order to enter the atmosphere and become visible. The meteors are like bugs that splatter on your auto windshield. You get many more on the front windshield as you drive down the highway than on the rear windshield, because the front windshield is driving into bugs and the rear windshield is driving away from bugs.

Watching a radiant sight: Meteor showers

Normally, only a few meteors per hour are visible — more after midnight than before and (for observers in the Northern Hemisphere) more in the fall than in the spring. But on certain occasions every year, you may see 10, 20, or even 50 or more meteors per hour in a dark, moonless sky far from city lights. Such an event is a *meteor shower,* when Earth passes through a great ring of billions of meteoroids that runs all the way around the orbit of the comet that shed them. (I discuss comets in detail later in this chapter.) Figure 4-1 illustrates the occurrence of a meteor shower.

The direction in space or place on the sky where a meteor shower seems to come from is called the *radiant.* The most popular meteor shower is the Perseids, which, at its peak, produces as many as 80 meteors per hour. (The Perseids get their name because they seem to streak across the sky from the direction of the constellation Perseus, the Hero, their radiant. Meteor showers are usually named for constellations or bright stars [such as Eta Aquarii] near their radiants.)

Figure 4-1: Earth's path crossing a belt of meteoroids creates a shower of meteors.

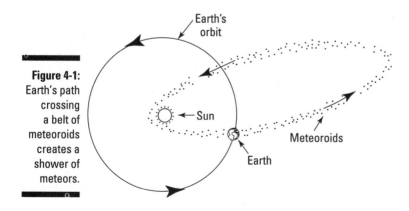

A few other meteor showers produce as many meteors as the Perseids, but fewer people take the time to observe them. The Perseids come on balmy nights in August, often perfect for skywatching, but the other leading meteor showers — the Geminids and Quadrantids — streak across the sky in December and January, respectively, when the weather is worse in the Northern Hemisphere and observers' ambitions are limited.

Table 4-1 lists the top annual meteor showers. The dates in the table are the nights when the showers usually reach their peak. Some showers go on for days, and others for weeks, raining down meteors at lower rates than the peak values. The Quadrantids may last for just one night or only a few hours.

Table 4-1	Top Annual Meteor Showers	
Shower Name	*Approximate Date*	*Meteor Rate (Per Hour)*
Quadrantids	Jan. 3–4	90
Lyrids	Apr. 21	15
Eta Aquarids	May 4–5	30
Delta Aquarids	July 28–29	25
Perseids	Aug. 12	80
Orionids	Oct. 21	20
Geminids	Dec. 13	100

The Quadrantids' radiant is in the northeast corner of the constellation Bootes, the Herdsman. The meteors are named for a constellation found on 19th-century star charts that astronomers no longer officially recognize. In addition to losing their namesake, the Quadrantids seem to have lost the comet that spawned them — their origin was a mystery until 2003, when astronomer Petrus Jenniskens found that an object named 2003 EH_1 may be their parent comet.

The Geminids are a meteor shower that seems to be associated with the orbit of an asteroid rather than a comet. However, the "asteroid" is probably a dead comet, which no longer puffs out gas and dust to form a head and tail. The object 2003 EH_1, the likely parent of the Quadrantids, may be a dead comet, too. (I discuss comets in the next section.)

The Leonids are an unusual meteor shower that occurs around November 17 every year, usually to no great effect. But every 33 years, many more meteors are present than usual, perhaps for several successive Novembers. Huge numbers of Leonids were seen in November 1966 and again in November 1999, 2000, 2001, and 2002, at least for brief times at some locations. The next great display likely will come in 2032.

You almost never see as many meteors per hour as I list in Table 4-1. The official meteor rates are defined for exceptional viewing conditions, which few people experience nowadays. But meteor showers vary from year to year, just like rainfall. Sometimes people do see as many Perseids as listed. On rare occasions, they see many more than expected. Such inconsistency is why keeping accurate records of the meteors that you count can be helpful to the scientific record.

Do you live south of the equator? If so, check out the list of meteor showers visible from the Southern Hemisphere on the website of the Royal Astronomical Society of New Zealand, at `www.rasnz.org.nz/Meteors/Meteors.htm`.

To track meteors, you need a watch, a notebook, a pen or pencil to record your observations, and a dim flashlight to see what you're writing.

The best light for astronomical observations is a red flashlight, which you can purchase or make from an ordinary flashlight by wrapping red transparent plastic around the bulb. Some astronomers paint the lamp with a thin coat of red nail polish. If you use a white light, you dazzle your eyes and make it impossible to see the fainter stars and meteors for 10 to 30 minutes, depending on the circumstances. Letting your vision adjust to the dark is called getting *dark adapted* and is a step you want to take every time you observe the night sky.

The best way to watch and count meteors is to recline on a lounge chair. (You can do pretty well just lying on a blanket with a pillow, but you're more likely to fall asleep in that position and miss the best part of the show.) Tilt your head so that you're looking slightly more than halfway up from the horizon to the zenith (see Figure 4-2) — the optimum direction for counting meteors. Take written notes, or use the NASA Meteor Counter App (which I describe shortly). And be sure you have a thermos of hot coffee, tea, or cocoa!

You don't have to face the radiant when you observe a meteor shower, although many people do. The meteors streak all over the sky, and their visible paths may begin and end far from the radiant. But you can visually extrapolate the meteors' paths back in the direction from which they seem to come, and the paths point back to the radiant. Identifying a radiant in that way is how you can tell a shower meteor from a sporadic one.

If you do face the radiant, however, you see some meteors that seem to have very short paths, even though they appear fairly bright. The paths appear short because the meteors are coming almost right at you. Fortunately, the shower meteoroids are microscopic and won't make it to the ground.

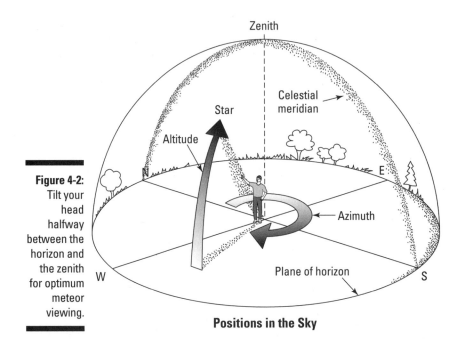

Figure 4-2:
Tilt your
head
halfway
between the
horizon and
the zenith
for optimum
meteor
viewing.

Positions in the Sky

Reporting meteors to NASA

You can help NASA by reporting the meteors that you see. Just download the NASA Meteor Counter App for your smartphone or tablet computer, and you're ready to go. You can access the free download and simple instructions on the web at www.meteorcounter.com. The Meteor Counter App displays a set of vertical rectangles that resemble the white keys on a piano. Each "key" stands for a different magnitude (brightness). Press the key that resembles the magnitude of the meteor that you see. You can even turn on a voice recorder function in the app to preserve comments such as "That meteor broke in two parts" or "It looks green." At intervals, your phone or tablet sends your observations to NASA, just as a space probe transmits data to Earth from time to time.

For help learning to judge magnitudes, consult these sources:

✔ **The British Astronomical Association Meteor Section** gives a short table of sky objects that you can compare with a meteor to judge its magnitude. Go to www.britastro.org/meteor and click the "Observing Notes" link; the table is near the end of the article.

✔ **The North American Meteor Network** offers free sky charts that you can download at www.namnmeteors.org/charts.html. Use them to compare stars of various magnitudes to the meteors that you observe. It's desirable to include in your observing notes an estimate of sky darkness. Just estimate the brightness of the faintest stars that you can see while observing. The darker the sky, the fainter the stars that are visible and the more meteors you can see.

Photographing meteors and meteor showers

You can photograph meteors on film or with a digital camera. But digital meteor photography requires a fairly expensive camera (point-and-shoot cameras don't work well, except in the rare case when you can catch a brilliant fireball) and a lot of trial-and-error experimenting until you get it right. Regardless of which method you use, certain guidelines apply:

✔ Observe from as dark a location as possible, away from urban lighting.

✔ Try meteor photography only when the Moon is below the horizon.

✔ Use a sturdy tripod so the camera doesn't shake during a time exposure.

✔ Use an ordinary lens or a wide-angle lens (the latter is preferable because you'll catch more meteors in a single shot) and set it on Infinity. Don't use a telephoto lens.

✔ Use a cable release to operate the camera shutter without shaking the camera.

✔ Point the camera about halfway up the sky from the horizon to the zenith, or a little higher, facing whichever direction has the least interfering sky glow from city or other lights.

When photographing meteors with a film camera, use these tips:

✔ Use color or black-and-white film with a speed of ISO 400.

✔ Use a lens that is rated f/5.6 or less and set it to the lowest f/number.

✔ Set the camera for a time exposure, preferably using the "bulb" mode, if your shutter has that capability. *Time exposure* means that you squeeze the cable release once to start the exposure and squeeze again to end it.

✔ Take a 10- or 15-minute exposure, then another, and repeat the process for an hour or two. When the film is developed, you may find that you've captured one or more meteor images. *Exception:* If a fireball passes through the sky area that the camera is viewing, end the exposure immediately and make a note of the time. You may have an important picture of a fireball track that won't benefit from further exposure time because the fireball is gone.

✔ When you have the film developed, tell the photo processor to "print all negatives." Film labs often avoid printing astronomical negatives, which may appear to be wasted or bad exposures to the nonastronomer.

Keep these pointers in mind when photographing meteors with a digital camera:

✔ Use a digital single lens reflex camera (DSLR) that is capable of time exposures.

✔ Spend some time making test exposures to determine what settings to use on that particular night. (The best settings vary depending on how bright the sky is.) Make several 10-second exposures, some 20-second exposures, and some 30-second exposures. You are trying to determine how long you can let an exposure last (the longer the better) without skylight overexposing the picture. You may need to repeat this series of time exposures for each of two or three ISO settings. (With a larger ISO setting, you can record fainter meteors, which means more meteors, but with the larger ISO setting, the sky overexposes sooner, so you can't expose for as long a time.) With experience, you should find the "sweet spot" of exposure time and ISO that works best with your lens at your location.

(continued)

(continued)

You can photograph sporadic meteors by following the preceding guidelines, but there aren't many sporadic meteors to catch on any given night. A meteor shower offers you the opportunity to snap more meteors, as long as the Moon isn't in the sky. With moonlight, you'll catch far fewer meteors, if any. When photographing a meteor shower, whether with film or digital cameras, take the photographs when the shower radiant (the constellation from which the meteor shower seems to come) is well above the horizon, preferably 40° or more. The horizon is at 0° altitude, and the zenith (overhead point) is 90° up, so the halfway point between them is at 45°; two-thirds of the way up is 60°, and so on.

Comets: *The Lowdown on Dirty Ice Balls*

Comets, great blobs of ice and dust that slowly track across the sky looking like fuzzy balls trailing gassy veils, are popular visitors from the depths of the solar system. They never fail to attract interest. Every 75 to 77 years, the best-known ice ball, Halley's Comet, returns to our neck of the woods. If you missed its appearance in 1986, try again in 2061! If you're impatient, you can see other interesting comets in the meantime. Often a less famous comet, such as Hale-Bopp in 1997, is much brighter than Halley's.

Many people confuse meteors and comets, but you can easily distinguish them by these criteria:

- ✔ A meteor lasts for seconds; a comet is visible for days, weeks, or even months.

- ✔ Meteors flash across the sky as they fall overhead, within 100 miles or so of the observer. Comets crawl across the sky at distances of many millions of miles.

- ✔ Meteors are common; comets that you can easily see with the naked eye come less than once a year, on average.

Astronomers believe that comets were born in the vicinity of the outer planets, starting near the orbit of Jupiter and extending well beyond Neptune. The comets near Jupiter and Saturn were gradually disturbed by the gravity of those mighty planets and flung far out into space, where they fill a huge, spherical region well beyond Pluto — the Oort Cloud — extending roughly 10,000 AU from the Sun. (I define the AU, or astronomical unit, a distance equal to about 93 million miles, in Chapter 1.) Other comets were ejected to or were formed and remain in the Kuiper Belt (see Chapter 9), a region that starts around the orbit of Neptune and continues to a distance of about 50 AU from the Sun, or about 10 AU beyond Pluto. Passing stars occasionally disturb these regions and send comets on new orbits, which may take them close to Earth and the Sun, where we can see them.

In the following sections, I discuss a comet's structure, famous comets throughout time, and methods you can use to spot a comet.

Making heads and tails of a comet's structure

A comet is a stuck-together mixture of ice, frozen gases (such as the ices of carbon monoxide and carbon dioxide), and solid particles — the dust or "dirt" shown in Figure 4-3. Historically, astronomers described comets as having a head and tail or tails, but with additional research, they've been able to clarify the nature of a comet's structure.

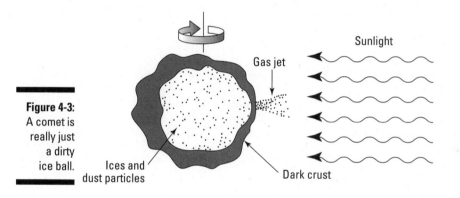

Figure 4-3: A comet is really just a dirty ice ball.

*(Labels in figure: Sunlight, Gas jet, Ices and dust particles, Dark crust)*

The nucleus

Astronomers initially named a bright point of light in the head of a comet the *nucleus.* Today we know that the nucleus is the true comet — the so-called dirty ice ball. The other features of a comet are just emanations that stem from the nucleus.

A comet far from the Sun is only the nucleus; it has no head or tail. The ice ball may be dozens of miles in diameter or just a mile or two. That size is pretty small by astronomical standards, and because the nucleus shines only by the reflected light of the Sun, a distant comet is faint and hard to find.

Images of Halley's nucleus from a European Space Agency probe that passed very close to it in 1986 show that the lumpy, spinning ice ball has a dark crust, like the tartufo dessert (balls of vanilla ice cream coated with chocolate) served in fancy restaurants. Comets aren't so tasty (I think), but they *are* real treats to the eye. Here and there on Halley's nucleus, the probe photographed plumes of gas and dust from geyserlike vents or holes sprayed into space where the Sun has barely warmed the surface. Some crust! And in 2004, NASA's Stardust probe got close-up images of the nucleus of Comet Wild-2. This nucleus seems to bear impact craters and is marked with what may be pinnacles made of ice. Those are the cold facts.

The coma

As a comet gets closer to the Sun, solar heat vaporizes more of the frozen gas and it spews out into space, blowing some dust out, too. The gas and dust form a hazy, shining cloud around the nucleus called the *coma* (a term derived from the Latin for "hair," not the common word for an unconscious state). Almost everyone confuses the coma with the head of the comet, but the head, properly speaking, consists of both the coma and the nucleus.

The glow from a comet's coma is partly the light of the Sun, reflected from millions of tiny dust particles, and partly emissions of faint light from atoms and molecules in the coma.

A tale of two tails

The dust and gas in a comet's coma are subject to disturbing forces that can give rise to a comet's tail(s): the dust tail and the plasma tail.

The pressure of sunlight pushes the dust particles in a direction opposite the Sun (see Figure 4-4), producing the comet's *dust tail*. The dust tail shines by the reflected light of the Sun and has these characteristics:

- ✔ A smooth, sometimes gently curved appearance
- ✔ A pale yellow color

The other type of comet tail is a *plasma tail* (also called an ion tail or a gas tail). Some of the gas in the coma becomes *ionized,* or electrically charged, when struck by ultraviolet light from the Sun. In that state, the gases are subject to the pressure of the *solar wind,* an invisible stream of electrons and protons that pours outward into space from the Sun (see Chapter 10). The solar wind pushes the electrified cometary gas out in a direction roughly opposite of the Sun, forming the comet's plasma tail. The plasma tail is like a wind sock: It shows astronomers who view the comet from a distance which way the solar wind is blowing at the comet's point in space.

In contrast to the dust tail, a comet's plasma tail has

- ✔ A stringy, sometimes twisted or even broken appearance
- ✔ A blue color

Now and then, a length of plasma tail breaks from the comet and flies off into space. The comet then forms a new plasma tail, much like a lizard that grows a new tail when it loses its first one. The tails of a comet can be millions to hundreds of millions of miles long.

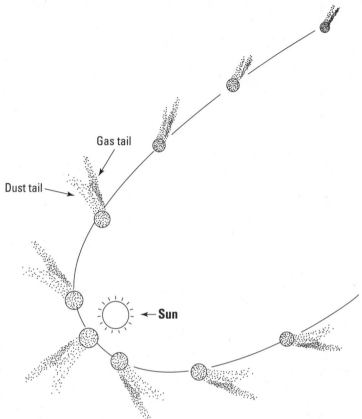

Gas tail

Dust tail

← **Sun**

Figure 4-4:
A comet's
tail points
away from
the Sun.

When a comet heads inward toward the Sun, its tail or tails stream behind it. When the comet rounds the Sun and heads back toward the outer solar system, the tail still points away from the Sun, so the comet now follows its tail. The comet behaves to the Sun as an old-time courtier did to his emperor: never turning his back on his master. The comet in Figure 4-4 could be going clockwise or counterclockwise, but either way, the tail always points away from the Sun.

The coma and tails of a comet are just a vanishing act. The gas and dust shed by the nucleus to form the coma and tails are lost to the comet forever — they just blow away. By the time the comet travels far beyond the orbit of Jupiter, where most comets come from, it consists of only a bare nucleus again. And the nucleus is a little smaller, due to the gas and dust that it sheds. The dust the comet loses may someday produce a meteor shower (which I cover earlier in this chapter), if it crosses Earth's orbit.

Coma again?

The first rule of viewing is, get out of town! Although a comet's nucleus may be just 5 or 10 miles in diameter, the coma that forms around it can reach tens of thousands or even hundreds of thousands of miles across. The gases are expanding, like a puff of smoke from a cigarette. As they thin out, they fade and become less visible. So the size of a comet's coma doesn't depend on just how much stuff the comet sheds; it also depends on the sensitivity of the human eye or the photographic film or digital camera sensor that you use to observe it. The apparent size of the coma also depends on the darkness of the sky in which you view it. A bright comet looks a lot smaller downtown than out in the country, where the skies are dark.

Halley's comet is a good example of the wasting-away process. Halley's nucleus decreases by at least a meter (39.37 inches, or slightly more than a yard) every 75 to 77 years when it passes near the Sun. The nucleus is only about 10 kilometers (10,000 meters) in diameter right now, so Halley's comet will survive only about 1,000 more orbits, or about 75,000 years. Dust shed by the famous comet causes two of the top annual meteor showers, the Eta Aquarids and the Orionids, which I list in Table 4-1.

Waiting for the "comets of the century"

Every few years, a comet is sufficiently bright and in such a good position in the sky that you can easily see it with the naked eye and with small binoculars. I can't tell you when such a comet is coming because the only comets whose returns astronomers can accurately predict in the near future are small ones that don't get very bright. Nearly all bright, exciting comets are discovered rather than predicted.

Halley's Comet is the only bright comet whose visits astronomers can accurately predict, but it doesn't come around often. Its appearance in 1910 was widely heralded, and everyone got a look. But an even brighter comet came the same year, the Great Comet of 1910, and no astronomer had predicted its arrival. All you can do is keep looking up. Monitor the astronomy magazines and the websites at the end of this section for reports of new comets, and follow the directions to view them. And with luck, you may be the first to spot and report a new comet, in which case the International Astronomical Union will name it after you.

Every five or ten years, a comet comes that is so bright astronomers hail it as "the comet of the century." People have short memories. But stay interested, and you may have a chance to see a fine comet:

✔ In 1965, Comet Ikeya-Seki was visible in broad daylight next to the Sun if you held up your thumb to block the bright solar disk. I'll never forget that sight — or my sunburned thumb.

✔ In 1976, Comet West was visible to the naked eye even in the night sky over downtown Los Angeles, one of the worst places I know of to see celestial objects.

✔ In 1983, Comet IRAS-Iraki-Alcock could be seen by the naked eye, actually moving in the night sky. (Most comets move so slowly across the constellations that you may have to wait an hour or more to notice any change in position.)

✔ In the 1990s, the bright comets Hyakutake and Hale-Bopp appeared out of the blue and were witnessed by millions of people worldwide.

✔ In 2007, Comet McNaught became the brightest comet since Ikeya-Seki in 1965 and, like that comet, was visible in the daytime.

The next great comet may come at any time. Keep looking, and you may even discover it!

Websites galore offer information on currently visible comets and photographs of them from amateur and professional astronomers. Most of the time, the current comets are too dim for any but advanced amateur telescopes. Check one or two of these good resources regularly to make sure you have the latest word:

✔ The Comet Chasing page gives detailed information on the visibility of comets each month for observers at various latitudes around the world, from 55° north to 30° south, at `www.cometchasing.skyhound.com`.

✔ Sky & Telescope's Comet page offers hints on how to observe and photograph comets, at `www.skyandtelescope.com/observing/objects/comets`.

✔ The Society for Popular Astronomy provides information and sky charts for comets visible from the United Kingdom (along with other sky phenomena). See its website news page at `www.popastro.com/news`.

✔ Southern Sky Watch, in Melbourne, Australia, provides information on comets visible from Down Under; check it out at `http://home.mira.net/~reynella/skywatch/ssky.htm`.

✔ The Heavens-Above website provides star charts for current comets; visit `www.heavens-above.com`.

Hunting for the great comet

Finding a comet isn't difficult, but finding your first one can take years and years. The famous comet hunter David Levy scanned the sky systematically for nine years before he found his first comet. Since then, he's found over 20 more.

The best telescope to use for comet searching is a *short focus* or *fast* telescope, meaning one whose catalogue specifications include a low f-number (like the f-number of a camera lens) — f/5.6 or, better yet, f/4. And you need to use a low-power eyepiece, such as 20x to 30x (see Chapter 3). The whole idea of the low f/number and the low magnification is to view as large an area of the sky as possible with your telescope. (It's called wide field observing.) The bright comets that you may be able to discover are few and far between, so you've got to look far and wide to catch them.

A relatively inexpensive telescope to start your comet hunt is the Orion ShortTube 80 Equatorial Refractor, with an 80-millimeter (3.1-inch) objective lens. It has good optics, an equatorial mount, and a tripod; its f/5.0 focal ratio and wide field eyepieces are just what you need for comet searches. This instrument lists for about $300 on the Orion Telescopes and Binoculars website, www. telescope.com. (For more about telescopes and mounts, check out Chapter 3, where Figure 3-4 shows a refractor with objective lens and eyepiece.)

You can search for unknown comets in two ways: the easy way and the systematic way. Read on to discover both techniques and for info on reporting a comet.

Locating comets the easy way

The easy way to search for comets is to make no extra effort at all. Just be on the lookout for fuzzy patches when you stare through your binoculars or telescope at stars or other objects in the night sky. Scan the sky for a fuzzy spot (as opposed to stars, which are sharp points of light if your binoculars are in focus). If you pinpoint a fuzzy area, check your star atlas to see whether anything at that location is *supposed* to look fuzzy, such as a nebula or a galaxy. If you find nothing like that on the atlas, you may have found a comet — but before you get too excited, wait a few hours and see if the possible comet moves against the pattern of adjacent stars. If the Sun rises or clouds move in the way and block your view, look again on the next night. If the object is indeed a comet, you'll notice that its position has changed with respect to the stars. And if the fuzz is bright enough, you may be able to spot a tail, which is a dead giveaway that you've found a comet.

Locating comets the systematic way

The systematic way to search for comets is based on the precept that you can find them most easily where they're brightest and where the sky is at its darkest. Comets closest to the Sun are the brightest, but the sky is darkest in directions far from the direction to the Sun.

Playing the space object name game

If you discover a comet, the International Astronomical Union will name it after you and possibly also after the next one or two people who independently report it.

If you discover a meteor, you don't have time to name it before it disappears. You can try shouting "John," but the name won't catch on and you may attract undue attention. The only meteors that get named are the spectacular ones seen by thousands of people. They get names such as the "Great Daylight Fireball of August 10, 1972," but no official procedure governs how meteors get named.

If you discover a meteorite, it's named for the town or other local area where you pick it up. The meteorite belongs to the owner of the land where you find it, and if you find it on U.S. government land, such as a national park or forest, it goes to the Smithsonian Institution.

If you discover an asteroid, you can recommend a name for the asteroid, but it can't be yours. (See Chapter 7 for more about asteroids.)

As a compromise between as far from the Sun and as close to the Sun as possible, look for comets in the east before dawn over the part of the sky that's

✔ At least 40° from the Sun (which is below the horizon)

✔ No more than 90° from the Sun

Remember that there are 360° all the way around the horizon, so 90° is one-quarter of the way around the sky.

A desktop planetarium program can help you map out the regions of the constellations that fit this bill for any given night of the year (see Chapter 2 for more about these programs). And, of course, you can look for comets in the west at dusk by following the same two rules about distance from the Sun. In my experience, the first few "comets" that you discover will be the contrails from jet airplanes, which catch the rays of the Sun at their high altitude, even though the Sun has set at your location.

Start at one corner of the sky area that you plan to check and slowly sweep the telescope across the area. Move the telescope slightly up or down and scan the next strip of sky in your search area. You can make every scan from left to right, or you can scan back and forth boustrophedonically (a term from classical times that refers to plowing a field with oxen; the oxen plow the first furrow in one direction and then come back across the field, plowing in the opposite direction).

You'll have an easier time impressing your friends by telling them of your boustrophedonic comet search project than by actually discovering a comet. It will give your ego a boost (unless your friends decide that you're plowed).

Reporting a comet

When you discover a comet, follow the directions on the website of the International Astronomical Union's Central Bureau for Astronomical Telegrams (which doesn't use telegrams anymore) and report it by e-mail. The site is www.cbat.eps.harvard.edu/index.html.

The bureau doesn't appreciate false alarms, so try to get a stargazing friend to check out your discovery before you spread the word. If the find checks out, you — as the amateur discoverer of a comet — may be eligible for a cash share of the Edgar Wilson Award, which is described on the Central Bureau website.

But even if you never discover a comet, and most astronomers never do, you can enjoy comets that others find.

Artificial Satellites: Enduring a Love–Hate Relationship

An artificial satellite is something that people build and launch into space, where it orbits Earth or another celestial body. The Earth-orbiting artificial satellites show us the weather, monitor El Niño, relay network television programs, and stand guard against intercontinental missile launches by hostile powers. And they can also be used for astronomy.

The Hubble Space Telescope is an artificial satellite, and astronomers love it. It gives unparalleled views of the stars and distant galaxies and lets you view the universe in ultraviolet and infrared light that's otherwise blocked by the thick layers of Earth's atmosphere. (As we go to press, NASA plans to keep the Hubble working for several more years, but by 2025, at the latest, it will be deorbited and will fall into the sea.)

But artificial satellites can also catch the rays of the setting Sun or even the Sun that has already set for observers at ground level. Reflecting the light of the Sun, they represent points of light that may move across the part of the sky where an astronomer is making a photograph of faint stars. Astronomers don't appreciate this interference. Worse yet, some artificial satellites broadcast at radio frequencies that interfere with the "big dish" and other radio antennas that astronomers use to receive the natural radio emanations from space. The celestial radio waves may have traveled for 5 billion years from a quasar, or they may have taken 5,000 years to reach us from another solar system in the Milky Way, possibly bearing a greeting from benevolent aliens who want to send us the cure for cancer. But just as the radio waves arrive, a blaring tone and strident modulations from a satellite passing over the observatory interfere with our reception. We may never know what the news is from the planet of the aliens.

So astronomers love satellites when they do something good for us and hate them when they interfere with our observations. But to make the best of a bad thing, amateur astronomers have become enthusiastic viewers and photographers of artificial satellites passing overhead.

Skywatching for artificial satellites

Hundreds of operating satellites are orbiting Earth, along with thousands of pieces of orbiting space junk — nonfunctional satellites, upper stages from satellite launch rockets, pieces of broken and even exploded satellites, and tiny paint flakes from satellites and rockets.

You may be able to glimpse the reflected light from any of the larger satellites and space junk, and powerful defense radar can track even very small pieces.

The best way to begin observing artificial satellites is to look for the big ones — such as NASA's International Space Station or the Hubble Space Telescope — and the bright, flashing ones (the dozens of Iridium communication satellites).

Looking for a big or bright artificial satellite can be reassuring to the beginning astronomer. Predictions of comets and meteor showers are sometimes mistaken, the comets usually seem fainter than you expect, and often you see fewer meteors than advertised. But artificial satellite viewing forecasts are usually right on. You can amaze your friends by taking them outside on a clear early evening, glancing at your watch, and saying "Ho hum, the International Space Station should be coming over about there (point in the right direction as you say this) in just a minute or two." And it will!

Want to know what to watch for? I've got you covered. Here are some characteristics you can pinpoint for both large and bright satellites:

✔ A big satellite such as the Hubble Space Telescope or the International Space Station generally appears in the evening as a point of light, moving steadily and noticeably from west to east in the western half of the sky. It moves much too slowly for you to mistake it for a meteor, and it moves much too fast for a comet. You can see it easily with the naked eye, so it can't be an asteroid — and, anyway, it moves much faster than an asteroid.

Sometimes you may confuse a high-altitude jet plane with a satellite. But take a look through your binoculars. If the object in view is an airplane, you should be able to distinguish running lights or even the silhouette of the plane against the dim illumination of the night sky. And when your location is quiet, you may be able to hear the plane. You can't hear a satellite.

✔ An Iridium satellite is a wholly different viewing situation: It usually appears as a moving streak of light that gets remarkably bright and then fades after several seconds. It moves much more slowly than a meteor. And an Iridium flare or flash is often brighter than Venus, second in brilliance only to the Moon in the night sky. The Sun, located below your horizon, reflects off one of the door-size, flat, aluminum antennas on the satellite to cause the flash of light. At star parties, people cheer when they spot an Iridium flare, just like when folks see a fireball. You can even see some Iridium flares in daylight.

And consider this: More than 60 Iridium satellites are in orbit. They interfere with astronomy, and astronomers want them to disappear, but at least the satellites have a "flare" for entertaining us.

Finding satellite viewing predictions

Some newspapers and television weather persons give daily or occasional forecasts for viewing satellites from the local area. You can get more detailed information whenever you want it by consulting these websites:

✔ For the International Space Station and the Hubble Space Telescope, *Sky & Telescope* offers a Satellite Tracker at www.skyandtelescope.com/observing/objects/javascript/3304316.html. When you use the Tracker, begin by changing the default settings to your own latitude, longitude, and time zone.

✔ For Iridium communication satellites, convenient forecasts are available from the Heavens-Above website at www.heavens-above.com. This site also provides predictions for daylight viewing of Iridium flares.

✔ Heavens-Above is also a good source place for Hubble Space Telescope viewing info. Click the "HST" link in the "Satellites" section to get a viewing schedule for your location.

You can also use a smartphone app to locate a bright satellite:

✔ If you have an iPhone, use Go Sat Watch.

✔ For Android phones, use Iridium Flares or Sky Junk.

After you succeed in viewing some bright artificial satellites, you can try photographing them. Follow the directions in the sidebar "Photographing meteors and meteor showers," earlier in this chapter. I don't recommend using point-and-shoot digital cameras for photographing meteors or most satellites. However, the International Space Station may be bright enough to catch with that kind of camera.

Part II
Going Once Around the Solar System

CARL FINALLY UNDERSTOOD WHY PLUTO WAS DEMOTED TO A DWARF PLANET

In this part . . .

*G*uess what? Men aren't from Mars, and women aren't from Venus. In fact, neither planet can support life as we know it. Venus is too hot, Mars is too cold, and neither has any liquid water that scientists know of.

This part explains what the planets of our solar system are really like. Did Mars ever support life? What about Jupiter's moon, Europa? New research and space missions are underway. Meanwhile, I tell you what scientists know now.

And if you've watched one of those "Oh no, a big asteroid is headed for Earth!" movies, you may be wondering whether to take a threat like that seriously. I include a chapter that explains asteroids and tells you the truth about the risk of their striking Earth.

Chapter 5

A Matched Pair: Earth and Its Moon

In This Chapter

▶ Seeing Earth as a planet

▶ Understanding Earth's time, seasons, and age

▶ Focusing on the Moon's phases and features

*P*eople often think of planets as objects up in the sky, like Jupiter and Mars. The ancient Greeks — and people for centuries thereafter — made a distinction between Earth, which they regarded as the center of the universe, and the planets. They thought of the planets as little lights in the sky that revolved around Earth.

Today we know better. Earth is a planet, too, not the center of the universe. It isn't even the center of our solar system; the Sun owns that title. The Moon orbits around Earth, along with hundreds of artificial satellites (see Chapter 4), and that's about it. Joining Earth in orbit around the Sun are seven other planets in the solar system, Pluto (called a dwarf planet), a large number of other moons, a belt of asteroids, millions of comets, and more. Nevertheless, as far as we know, life exists in our solar system only on Earth.

Earth has fallen from its exalted place in human thought as the center of the universe to its true, but still significant, status: our home planet. And no other place in the solar system is quite like home.

Earth is what astronomers call a *terrestrial* planet — a kind of circular definition because *terrestrial* means "earthly." But the scientific meaning is a planet made of rock that orbits the Sun. The four planets closest to the Sun are our solar system's terrestrial planets: Mercury, Venus, Earth, and Mars, in order of distance from the Sun.

Some people consider Earth's Moon a terrestrial planet and regard the Earth–Moon system as a double planet. For aliens seeking to visit us, that distinction may help: "Just head for the yellow-white star in Sector 49,832 of the Orion Arm, in the Milky Way, and home in on the third rock from that Sun; it's a double planet and easy to spot."

Putting Earth under the Astronomical Microscope

Earth is unique among the known planets. In the following sections, I tell why, briefly summarizing some of its main characteristics and how they play into astronomical topics like time and the seasons. And in case you forgot what it looks like, you can check out a nice NASA photo in the color section, which shows Earth and the Moon together.

One of a kind: Earth's unique characteristics

What's so special about Earth? For starters, we inhabit the only planet we know of with these characteristics:

- **Liquid water at the surface:** Earth has lakes, rivers, and oceans, unlike any other known planet. Unfortunately, it has tsunamis and hurricanes, too. The oceans cover 70 percent of the surface of Earth.

- **Plentiful amounts of oxygen in the air:** The air on Earth contains 21 percent oxygen; no other planet has more than a trace of oxygen in its present atmosphere, as far as we know. (Most of the Earth's atmosphere, about 78 percent, is nitrogen.)

- **Plate tectonics (also known as *continental drift*):** Earth's crust is composed of huge moving plates of rock; where plates collide, earthquakes occur and new mountains rise. New crust emerges at the midocean ridges, deep beneath the sea, causing the seafloor to spread. (To find out about an interesting seafloor property, see the sidebar "Earth's seafloor and its magnetic properties," later in this chapter.)

- **Active volcanoes:** Hot molten rock, welling up from deep beneath the surface, forms huge volcanic landforms such as the Hawaiian Islands. Volcanoes erupt somewhere on Earth every day.

- **Life, intelligent or otherwise:** I'll let you be the judge on intelligence, but from one-celled amoebas, bacteria, and viruses to flowers and trees, fish and fowl, and insects and mammals, Earth has life in abundance.

Researchers are investigating tantalizing indications that Mars and Venus may once have shared some of these traits with Earth (see Chapter 6). But as far as we know, they don't have life now, and we don't have proof that they ever did.

Scientists believe that the presence of liquid water on the surface of Earth is one of the main reasons life flourishes here. You can easily imagine advanced life forms on other worlds — you see them on television and in the movies. But the images you see are all imaginary. Scientists don't have convincing evidence for any life, past or present, anywhere but on Earth.

Spheres of influence: Earth's distinct regions

Figure 5-1 shows four views of Earth as seen from space. Earth's patterns of land, sea, and clouds are clearly visible.

Figure 5-1:
Four views showing the changing face of Earth.

Courtesy of NASA

Enjoying the northern lights

The aurora is one of the most beautiful sights of the night sky and, for many people, a rare one. Depending on whether you live in the Northern Hemisphere or the Southern Hemisphere, you can see the *aurora borealis* (northern lights) or the *aurora australis* (southern lights), respectively.

Auroras appear when streams of electrons from Earth's magnetosphere rain down on the atmosphere below, stimulating oxygen and other atoms to shine. The eerie glow in the night sky may remain stationary for minutes to hours or constantly change (making it hard for a novice observer to identify). It can shimmer, pulsate, or flash around the sky. The aurora appears in many forms; here are a few of the most common:

- ✔ **Glow:** The simplest aurora. It looks like thin clouds are reflecting moonlight or city lights. But you don't see clouds — just the eerie light of an aurora.

- ✔ **Arc:** Shaped like a rainbow, but with no sunlight to make one. A steady or pulsating green arc is the most common type of arc, but sometimes dim red arcs appear.

- ✔ **Curtain:** Also called drapery. This spectacular aurora resembles a billowing curtain at a theater, but nature is the star of the show.

- ✔ **Rays:** One or more long, thin bright lines in the sky, appearing like faint beams from the heavens.

- ✔ **Corona:** High overhead, a crown in the sky, with rays emanating in every direction.

Auroras occur constantly in two geographical bands around Earth at high northern and southern latitudes. Folks who live beneath these two *auroral ovals* see auroras every night. But you may encounter big exceptions: When a great disturbance in the solar wind (see Chapter 10) strikes the magnetosphere, the ovals move toward the equator. People in the *auroral zones* (the lands beneath the ovals) may miss their aurora, but skygazers toward the equator who rarely see them are treated to a great show. The most likely times to see bright auroras outside the auroral zones are the first few years after the peak of the sunspot cycle, so keep your eyes open for auroras in 2013 and the next few years. If you don't want to wait for aurora to come to you, visit a high-latitude location, where you're near the northern auroral oval and can see aurora on almost any clear, dark night. Famous aurora-viewing places include

- ✔ **Fairbanks, Alaska** (`http://fair banks-alaska.com/northern-lights-alaska.htm`)

- ✔ **Yellowknife, Canada,** capital of the Northwest Territories, where you can relax in the Aurora Village's "reclining, heated outdoor viewing seats or teepees" and enjoy a sky show (`www.auroravillage.com`)

- ✔ **Tromsø, Norway** and vicinity, where you watch aurora from just outside your hotel, experience it on horseback in a fjord, take in the northern lights on a self-guided snowshoe trek, ride by reindeer-drawn sled to a remote overnight viewing spot, or have a similar adventure by dogsled (`www.visit tromso.no/de/Activities/ Aurora-Experiences`)

The good news about these northern locales is that you'll have great auroral views (weather permitting). The bad news is that the best time to see aurora in these cold places is from December to March, the coldest part of the year. At other times, the nights are shorter, so viewing prospects are poorer.

If you live in the Northern Hemisphere and want to look for aurora, check the daily aurora borealis forecast at the University of Alaska Geophysical Institute website, at `www. gi.alaska.edu/AuroraForecast.`

Better yet, get the forecast on your smartphone; just download the necessary app from that site. In Great Britain, you can subscribe to AuroraWatch UK to receive aurora alerts by e-mail or via social media. Sign up at `http://aurorawatch.lancs.ac.uk`.

If the aurora isn't visible at your location, you can still watch live images of it (except during Northern Hemisphere summer) from the AuroraMAX Camera of the Canadian Space Agency. Just visit `www.asc-csa.gc.ca/eng/astronomy/auroramax/connect.asp`.

Scientists classify the regions of Earth into these categories:

- ✔ The *lithosphere:* The rocky regions of our planet
- ✔ The *hydrosphere:* The water in the oceans, lakes, and elsewhere on Earth
- ✔ The *cryosphere:* The frozen regions — notably, the Antarctic and Greenland ice caps
- ✔ The *atmosphere:* The air from ground level up to hundreds of miles
- ✔ The *biosphere:* All living things on Earth — on land, in the air and water, and underground

So you're part of the biosphere that lives on the lithosphere, drinks from the hydrosphere, and breathes the atmosphere. (You can also visit the cryosphere.) I don't know anywhere else in space you can do all that.

In addition to the regions I describe in the previous list, one more important part of our planet is the *magnetosphere,* which plays a vital part in protecting Earth from many of the dangerous emanations from the Sun (see Chapter 10). Sometimes called Earth's radiation belts (or the Van Allen radiation belts, named for James Van Allen, a U.S. physicist who discovered them with America's first artificial satellite, Explorer 1), the magnetosphere consists of electrically charged particles — mostly electrons and protons — that bounce back and forth above Earth, trapped in its magnetic field.

Occasionally, some of the electrons escape and rain down on Earth's atmosphere below, striking atoms and molecules and making them glow. That glow is the aurora (see the sidebar "Enjoying the northern lights," earlier in this chapter, for more about viewing auroras).

The solid surface of Earth — the part you stand on — is the crust. Beneath the crust are the mantle and the core. The core is largely iron and nickel and is very hot, reaching about 12,000°F (about 7,000°C) at the center. The core is layered, too: The outer core is in a molten state, but the inner core is solid.

Earth's seafloor and its magnetic properties

According to geophysical surveys, patterns of magnetized rock exist in the seafloor on either side of midocean ridges. The rock became magnetized as it cooled from the molten state, trapping and "freezing in" some of Earth's magnetic field that pervaded it as the rock solidified. So the seafloor rock resembles a magnet, with a magnetic field that has strength and direction. After the rock solidified, its magnetic field could change no longer, and it became a fossil magnetic field. It's like a fossil dinosaur that remains forever in the shape it had at death.

The patterns discovered near the midocean ridges consist of stripes of magnetized rock, hundreds of miles long, that parallel the ridges and alternate in polarity. One stripe has a north magnetic polarity, like the end of a bar magnet that attracts a north-seeking compass needle; the next stripe has the opposite polarity; and so on.

The alternating stripes of oppositely magnetized rock are the result of the new rock emerging from the midocean ridges, cooling and magnetizing, and moving away from the ridges as even newer rock pushes it along. The oppositely magnetized stripes show that the geomagnetic field itself reverses direction from time to time, like a bar magnet that you turn 180° at intervals — except that the intervals for the geomagnetic field range from about a thousand to a million years.

An unknown process causes the geomagnetic field, generated deep in Earth's core, to reverse every so often. That effect is preserved in the fossil magnetic fields of the rock at the seafloor and in rock on the continents that previously lay beneath the sea.

Why mention all this seafloor stuff in a book on astronomy? Because this unique property of Earth may correspond to a phenomenon discovered on Mars. As scientists consider the evidence gathered on the various terrestrial planets, including Earth, we find similarities and differences that help us understand them better. Such research is called *comparative planetology,* and I cover it in more detail in the descriptions of Mars and Venus in Chapter 6.

The extremely high pressure of the overlying layers makes the hot iron in the inner core solidify. As Earth cools over millions of years in the future, the solid part at the center increases in size at the expense of the surrounding molten core, like an ice cube growing as the surrounding liquid cools.

Earth's core is far below our digging range, but it produces an effect that anyone can observe at the surface. Moving streams of molten iron in the outer core generate a magnetic field that reaches out through the whole planet and far into space, called the *geomagnetic field.*

The geomagnetic field

✔ Makes the needle of a compass point toward north (or south)

✔ Provides an invisible guidance system for homing pigeons, some migratory birds, and even some ocean-dwelling bacteria

✔ Forms the magnetosphere far above Earth

✔ Shields Earth from incoming electrically charged particles from space, such as the solar wind and many cosmic rays (high-speed, high-energy particles that come from explosions on the Sun and from distant points in space)

The geomagnetic field is a global planetary magnetic field, meaning that it extends above all parts of Earth and is continuously being generated. Mars, Venus, and our Moon all lack a global magnetic field like Earth's, and this key difference gives scientists information about the cores of those objects. For more on the lunar core, see the section "Quite an impact: Considering a theory about the Moon's origin," later in this chapter.

Examining Earth's Time, Seasons, and Age

It may be hard to believe because you can't walk 5 feet anymore without having access to a clock, but the rotation of Earth was the original basis of our system of measuring time, and we now know that the orbital motion of Earth and the tilt of its axis produce the seasons. Our seasonal ring-around-the-rosy (or yellowy, in this case) Sun dance has been going on for a long time; Earth is about 4.6 billion years old.

Orbiting for all time

Nowadays, scientists have atomic clocks that measure time with great precision. But originally and until modern times, our system of time was based on the rotation of Earth.

Knowing how time flies

Earth rotates once on its axis in 24 hours. It turns from west to east (which is counterclockwise as seen from space above the North Pole). The length of the day, 24 hours, is the average time it takes for the Sun to rise and set and rise again. This process is called *mean solar time* and is equivalent to the standard time on your watch.

The length of the day, therefore, is 24 hours of mean solar time. And a year consists of approximately 365 days, the time that it takes Earth to make one complete orbit around the Sun.

Because Earth moves around the Sun, the time when you see the Sun rise depends on both the rotation of Earth and Earth's orbital motion.

Earth turns once in 23 hours, 56 minutes, and 4 seconds with respect to the stars. That amount of time is called the *sidereal day.* (*Sidereal* means "pertaining to the stars.") Notice that the difference between 24 hours and 23 hours, 56 minutes, and 4 seconds is 3 minutes and 56 seconds, which is just about ⅟₃₆₅ of a day. The difference is no coincidence: It happens because, during a day, Earth moves through ⅟₃₆₅ of its orbit around the Sun.

Astronomers used to depend on special clocks called sidereal clocks, which measured sidereal time by registering 24 sidereal hours during an interval of 23 hours, 56 minutes, and 4 seconds of mean solar time. The sidereal hours, minutes, and seconds are all slightly shorter than the corresponding units of solar time. Using sidereal clocks enabled astronomers to keep track of the stars in order to point telescopes correctly. But we don't need to do that anymore. Computer programs that point telescopes or that picture the sky on a desktop planetarium or smartphone, as I describe in Chapter 2, do the mathematics for you, so you can simply use the standard time at your location to figure out where different stars and constellations appear in the sky.

On the other hand, astronomers still adhere to the custom of reporting astronomical observations on a common system called *Universal Time (UT)* or *Greenwich Mean Time.* UT is simply the standard time at Greenwich, England. If you live in North America, the standard time at your location is always earlier than the time in Greenwich. For example, in New York City, the Sun rises about 5 hours after it rises in Greenwich. When the clock strikes 6 a.m. in Greenwich, clocks in New York are turning to 1 a.m.

A more precisely defined time, *Coordinated Universal Time,* or *UTC,* which is identical to UT for all practical purposes, is the official international standard.

Looking out for leap seconds

Earth takes 365¼ days to go around the Sun, but a normal calendar year is just 365 days. That's why we add a day, February 29, every fourth year. We call that fourth year *leap year,* and the added day helps to keep Earth and the calendar in sync.

There's also a sync problem between the rotation of Earth and the length of a calendar day: Every now and then, Earth turns a wee bit slower than usual, perhaps due to a large weather system such as El Niño. These discrepancies accumulate and, before you know it, Earth is turning is out of whack with time kept by ultraprecise atomic clocks. On such occasions, international authorities declare a *leap second,* which is added to the UTC. When that little adjustment is made, engineers worldwide hold their breath and cross their fingers, hoping that it doesn't mess up GPS, air traffic control, and other critical, time-dependent systems. The United States and some other countries want to abolish leap seconds, but the United Kingdom and others want to keep them. A decision on the future of the leap second is expected in 2015.

Finding the right time

In the United States, the U.S. Naval Observatory (USNO) in Washington, D.C., is in charge of time. You can get the UTC any time you want it from the USNO Master Clock web page at `http://tycho.usno.navy.mil/what.html`. The USNO site also has a place where you can find the Local Apparent Sidereal Time at your location. The *Local Apparent Sidereal Time* equals the right ascension (see Chapter 1) of the stars on your meridian — the imaginary line from the zenith to the south point on the horizon. A star is best placed for observation when it's on the meridian.

To determine the standard time zone that applies at just about any other place in the world and to convert it to Universal Time, consult the World Time Zone Map of Her Majesty's Nautical Almanac Office at `http://aa.usno.navy.mil/faq/docs/world_tzones.php`.

Generally, daylight saving time (called British Summer Time in the United Kingdom) is an hour later than standard time at the same geographic location. But not all places observe daylight saving time. For example, Arizona, which gets plenty of sunshine year-round, never goes on daylight saving time.

Tilting toward the seasons

Teaching students about the cause of the seasons is about the most frustrating task of any astronomy professor. No matter how carefully the professor explains that the seasons have nothing to do with how far we are from the Sun, many students don't grasp it. Even surveys taken at Harvard University graduation show that bright college graduates think that summer occurs when Earth is closest to the Sun and winter occurs when Earth is farthest from the Sun.

Students forget that when summer comes in the Northern Hemisphere, the South experiences winter. And when Australians are surfing in the summer, people in the United States are wearing their winter coats. But Australia and the United States are on the same planet. Earth can't be farthest from the Sun and closest to the Sun at the same time. Earth is a planet, not a magician.

The actual cause of the seasons is the tilt of Earth's axis (see Figure 5-2). The *axis,* the line through the North and South poles, isn't perpendicular to the plane of Earth's orbit around the Sun. Actually, the axis is tilted by 23½° from the perpendicular to the orbital plane. The axis points north to a place among the stars — in fact, near the North Star (at least, in the short term; the axis slowly changes its pointing direction, so the North Star in one era will no longer be the North Star in the far future).

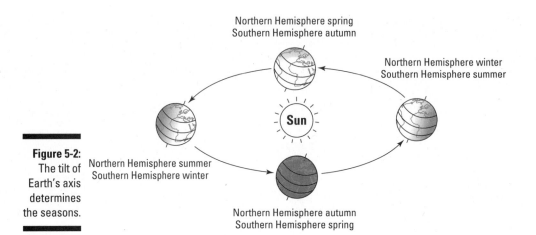

Northern Hemisphere spring
Southern Hemisphere autumn

Northern Hemisphere winter
Southern Hemisphere summer

Sun

Northern Hemisphere summer
Southern Hemisphere winter

Northern Hemisphere autumn
Southern Hemisphere spring

Figure 5-2:
The tilt of
Earth's axis
determines
the seasons.

At present, the North Star, also called Polaris, is the star Alpha Ursae Minoris, located in the Little Dipper asterism of the Little Bear constellation, Ursa Minor. If you get lost at night and want to "bear" north, set your sights on the Little Dipper (see Chapter 3 for more about finding Polaris).

The axis of Earth points "up" through the North Pole and "down" through the South Pole. When Earth is on one side of its orbit, the axis pointing up also points roughly toward the Sun so that the Sun looms high in the sky at noon in the Northern Hemisphere. Six months later, the axis points up and roughly away from the Sun. Actually, the axis always points in the same direction in space, but Earth has now moved to the opposite side of the Sun.

Summer occurs in the Northern Hemisphere when the axis pointing up through the North Pole points roughly at the Sun. When that happens, the Sun at noon is higher in the sky than at other seasons of the year, so it shines more directly on the Northern Hemisphere and provides more heat. At the same time, the axis pointing down through the South Pole points away from the Sun, so it shines lower in the sky at noon than at any other season of the year, creating less direct sunlight — and, thus, winter in Australia.

We enjoy more hours of sunlight in the summer because the Sun is higher in the sky. It takes longer for the Sun to rise to that height, and it takes longer to set.

As we orbit the Sun, it seems to move through the sky, following a circle called the *ecliptic,* which I mention in Chapter 3. The ecliptic is tilted with respect to the equator by exactly the same angle as Earth's tilt on its axis: 23½°. Here are some key events in the Sun's annual journey around the ecliptic, as experienced in the Northern Hemisphere:

- ✔ **Vernal equinox:** On the first day of spring, the Sun crosses from "below" (south) the equator to "above" (north).

- ✔ **Summer solstice:** The Sun reaches the farthest point north on the ecliptic.

- ✔ **Autumnal equinox:** The Sun crosses the equator going back down south, and fall begins.

- ✔ **Winter solstice:** The Sun gets as far south as possible on the ecliptic.

In the Northern Hemisphere, the summer solstice is the day with the most hours of sunlight during the year because the Sun attains its highest position in the sky — it takes the longest time to reach that height and come back down to the horizon again. By the same token, the winter solstice in the Northern Hemisphere is the day with the shortest amount of daylight during the year.

And that's the long and the short of time and seasons.

Estimating Earth's age

Measuring radioactivity is the only accurate way we have to date very old things on Earth or in the solar system. Some elements, such as uranium, have unstable forms called *radioactive isotopes.* A radioactive isotope turns into another isotope of the same element, or into a different element, at a rate determined by the *half-life* of the radioactive substance. If the half-life is 1 million years, for example, half of the radioactive isotope that was originally present will have turned into another substance (called the *daughter isotope*) by the time 1 million years has elapsed, leaving half still radioactive. And half of the remaining half turns into daughter isotope atoms in another million years. So after 2 million years, only 25 percent of the original radioactive isotope atoms still exists. After 3 million years, only 12½ percent remains. And so on.

When the original radioactive isotope atoms, called the *parent atoms,* and the daughter atoms are trapped together in a piece of rock or metal, such as a meteorite, scientists can count the atoms' respective numbers to determine how old the rock is in a process called *radioactive dating.*

Scientists have used radioactive dating to determine that the oldest rocks on Earth are about 3.8 billion years old. However, Earth is undoubtedly much older than that. Erosion, mountain building, and *volcanism* (the eruption of molten rock from within Earth, including the formation of new volcanoes) constantly destroy the rocks at the surface, so the original surface rocks of Earth are long gone.

Meteorites, however, yield radioactive dates as old as 4.6 billion years. Meteorites are considered debris from asteroids, and asteroids are thought to be debris from the very early solar system, when the planets first formed (see Chapter 7 for more about asteroids).

So scientists think that Earth and other planets are about 4.6 billion years old. Earth's Moon, however, is a little younger, as I explain in the next section.

Making Sense of the Moon

The Moon is 2,160 miles (3,476 kilometers) in diameter, slightly more than a quarter of the diameter of Earth. The Moon has no meaningful atmosphere — just a trace of hydrogen, helium, neon, and argon atoms, along with other traces in even lesser quantities. It's all, or mostly all, made of solid rock (see Figure 5-3); some experts think it may have a small molten iron core. Its mass is only $\frac{1}{81}$ the mass of Earth, and its density is about 3.3 times the density of water, which is noticeably less than the density of Earth (5.5 times the density of water).

The following sections give you the lowdown on the Moon's phases, lunar eclipses, and its geology (including handy tips for viewing a variety of lunar features). I also share a theory about the Moon's origin.

Figure 5-3:
The Moon is made of rocks and rilles, craters, and dried lava plains — not a speck of cheese in sight.

Courtesy of NASA

Get ready to howl: Identifying phases of the Moon

Except during a lunar eclipse (see the next section), half the Moon is always in sunlight and half is always in night. But contrary to popular belief, these light and dark hemispheres don't correspond to the lunar near side and the lunar far side. Those sides are the hemispheres that point toward and away from Earth, which are always the same. The lunar halves in sunlight and in night are the hemispheres that face toward and away from the Sun. And they always change as the Moon moves around Earth (see Figure 5-4).

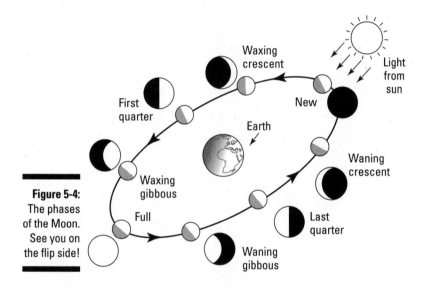

Figure 5-4:
The phases
of the Moon.
See you on
the flip side!

New Moon is the beginning of the monthly lunar cycle, or *lunation.* At this time, the near side faces away from the Sun, making it the dark side. A few hours or days later, the Moon is a new crescent, or *waxing crescent,* meaning a crescent Moon whose bright area is getting larger. This phase happens as the Moon moves away from the Sun–Earth line while orbiting Earth. Fully half of the Moon is always lit up, facing the Sun, but during a crescent Moon, we can't see most of this illuminated area that faces away from Earth.

As the Moon moves around its orbit, it reaches a point where the Earth–Moon line is at right angles to the Earth–Sun line. At this stage, we see a *half Moon,* which astronomers call a *quarter Moon.*

How can a half equal a quarter? It can't if you're trying to make change, but an astronomer can easily make it work. Half of the lunar near side — the part facing Earth — is lit up, so people call it a half Moon. But the illuminated portion of the Moon that we see is only half the bright hemisphere that faces the Sun, and half of a half is a quarter. Bet your friends that a quarter can be a half. You'll win, and you can pocket the change.

When the illuminated part of the Moon that we can see grows larger than the quarter (half) Moon but is still smaller than the full Moon, astronomers call it a *waxing gibbous Moon.*

When the Moon is on the far side of its orbit, opposite the Sun in the sky, the lunar hemisphere that faces Earth is fully lit, creating a *full Moon.* As the Moon continues around its orbit, the illuminated portion gets smaller and the Moon becomes gibbous again, less than full and more than a quarter Moon (a *waning gibbous Moon*). Soon the Moon appears as a quarter Moon again, called *last quarter.* As the Moon nears the line between Earth and the Sun, it becomes a *waning crescent Moon.* Soon it becomes a new Moon again, and the cycle of phases starts over.

The period of time over which the lunar phases change from new Moon to the next new Moon is the *synodic month,* which averages 29 days, 12 hours, and 44 minutes.

People often ask why an eclipse of the Sun doesn't occur every month at new Moon. The reason is that Earth, the Moon, and the Sun usually aren't all exactly on a line at a new Moon. When they are all in line at new Moon, an eclipse of the Sun results (see Chapter 10). When the three bodies are all on a line at a full Moon, we witness an eclipse of the Moon.

Earth has phases, too! To see them, however, you need to head into space and look back at Earth from a distance. When folks on Earth see a beautiful full Moon, an observer standing on the near side of the Moon would enjoy a "new Earth," and when earthlings experience a new Moon, viewers on the Moon would see a "full Earth."

In the shadows: Watching lunar eclipses

A lunar eclipse occurs when a full Moon is exactly on the line from the Sun to Earth. The Moon is then in Earth's shadow, or the *umbra.* You can safely look at a lunar eclipse, as long as you don't bump into something in the dark or stand in the road.

During a total eclipse of the Moon, you can still see the Moon, although it's immersed in Earth's shadow (see Figure 5-5). No direct sunlight falls on it, but some light from the Sun gets bent around the edges of Earth's atmosphere (as visible from the Moon) and falls on the Moon. The sunlight gets strongly filtered as it passes through our atmosphere, so mostly the red and orange light gets through. This *earthshine* effect differs from one lunar eclipse to the next, depending on meteorological conditions and the clouds in Earth's atmosphere. The totally eclipsed Moon, therefore, can look a dull orange, an even duller red, or a very dark red. Sometimes you can barely make out the eclipsed Moon at all.

The dates for upcoming total lunar eclipses through the year 2025 are

> *April 15, 2014*
>
> *October 8, 2014*
>
> *April 4, 2015*
>
> *September 28, 2015*
>
> *January 31, 2018*
>
> *July 27, 2018*
>
> *January 21, 2019*
>
> *May 26, 2021*
>
> *May 16, 2022*
>
> *November 8, 2022*
>
> *March 14, 2025*
>
> *September 7, 2025*

To prepare for the upcoming eclipses, you can find plenty of information on the exact times and on the part of Earth where the eclipse will be visible. Take a look in *Astronomy* or *Sky & Telescope* magazines and on their websites (www.astronomy.com and www.skyandtelescope.com) as the eclipse dates approach.

Total eclipses of the Moon are as common as total eclipses of the Sun, but you see them more often at any given place because a total eclipse of the Sun is visible only along a narrow band on Earth called the *path of totality*. But when Earth's shadow falls on the Moon, you can see the eclipsed Moon from all over half of Earth, wherever night has fallen.

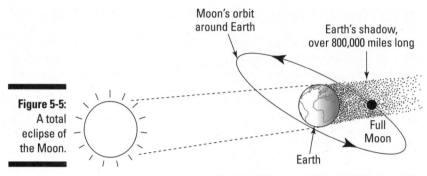

Moon's orbit
around Earth

Earth's shadow,
over 800,000 miles long

Full
Moon

Earth

Dinah L. Moché/Astronomy: A Self-Teaching Guide, Seventh Edition

Partial eclipses aren't quite as interesting. During a *partial eclipse,* only some of the full Moon falls within Earth's shadow. The Moon just appears to be at a different phase. If you don't know that an eclipse is in progress — or that the full Moon phase is underway — you aren't aware that a unique astronomical event is occurring. You may simply pass it off as a quarter or crescent Moon. But if you keep looking for an hour or so, you can see the full Moon come out from Earth's shadow.

Hard rock: Surveying lunar geology

The entire Moon is pockmarked with craters of every size, from microscopic pits to basins hundreds of miles in diameter. The largest is the South Pole–Aitken Basin, which is about 1,600 miles (2,600 kilometers) across. Objects (asteroids, meteoroids, and comets) that struck the Moon — very long ago, for the most part — caused these craters. The microscopic craters, which scientists have found on rocks brought back by astronauts from the surface of the Moon, are caused by micrometeorites — tiny rock particles flying through space. All the craters and basins are known collectively as *impact craters,* to distinguish them from volcanic craters.

The Moon has experienced volcanism, but it took a form different from Earth's. The Moon has no *volcanoes,* or large volcanic mountains with craters at the top. But it does have small volcanic domes, or round-topped hills like ones that occur in some volcanic regions on Earth. In addition, sinuous channels on the lunar surface (called *rilles)* appear to be lava tubes, also a common landform in volcanic areas on Earth (such as Lava Beds National Monument in northern California). Most notably, the Moon has huge lava plains that fill the bottoms of the large impact basins. These lava plains are called *maria,* the Latin word for "seas." (When you look up and see the Man in the Moon, the dark areas that make up some of his features are the maria.)

Some early scientists thought that the maria could be oceans. But if they were oceans, you could see bright reflections of the Sun from them, just as you do when you look down at the sea from an airplane during the day. The larger, bright areas in the Man in the Moon are the *lunar highlands,* which are heavily cratered areas. The maria have craters, too, but fewer craters per square mile than the highlands, which means that the maria are younger. Huge impacts created the basins where the maria are located. These impacts obliterated preexisting craters. Later the basins filled with lava from below, wiping out any new craters that had formed after the huge impacts. All the craters you can see in the maria now are from impacts that happened after the lava froze.

Moon ice? That's nice

A so-called lunar soil, consisting of fine rock dust, covers the surface of the Moon. It comes from countless impacts of meteoroids and asteroids that have struck the Moon for ages, making craters and pulverizing rock. Frozen molecules of water are stuck to the dust particles in many cases, especially in the bottoms of craters near the poles of the Moon. In the vicinity of those craters, the Sun is never high in the sky, and the crater bottoms are shadowed by the crater rims. These areas are the coldest places on the Moon. The temperature in at least one south pole crater gets below –400°F. But don't plan on scooping up Moon dust to make a refreshing glass of lunar ice water. Along with the water molecules, silver and mercury atoms are frozen to the dust, and you don't want them sliding down your throat.

Ready to observe the Moon's near side and find out the scoop on the Moon's far side? Check out the following sections.

Observing the near side

The Moon is one of the most rewarding objects to observe. You can see it when the sky is hazy or partly cloudy, and at times it's visible during the day. You can see craters with even the smallest telescopes. And with a high-quality small telescope, you can enjoy hundreds, and maybe thousands, of lunar features, such as impact craters, maria, lunar highlands, rilles, and other features, including the following

- **Central peaks:** Mountains of rubble thrown up in the rebound of the lunar surface from the effects of a powerful impact. Central peaks are found in some, but not all, impact craters.

- **Lunar mountains:** The rims of large craters or impact basins, which may have been partly destroyed by subsequent impacts, leaving parts of their walls standing alone like a range of mountains, although not the type of mountain you see on Earth.

- **Rays:** Bright lines formed by powdery debris thrown out from some impacts. They extend radially outward from young, bright impact craters, such as Tycho and Copernicus (see Figure 5-6).

Figure 5-6:
A close-up
view of the
lunar crater
Copernicus
from the
Hubble
Space
Telescope.

Courtesy of John Caldwell (York University, Ontario), Alex Storrs (STSci), and NASA

If you want to be able to distinguish one crater, rille, or lunar mountain range from another as you look through your telescope, you need a Moon map or a set of lunar charts. These inexpensive items are available from astronomy and other scientific hobby supply houses and sometimes from map stores. Here are two good sources for these maps:

  ✔ Orion Telescopes & Binoculars (www.telescope.com) sells maps and helpful guidebooks for lunar observing.

  ✔ The website www.skyandtelescope.com offers Moon maps drawn by lunar cartographer Antonín Rükl.

Remember, these maps and charts show only one side of the Moon: the lunar near side. You don't need a map of the far side of the Moon because you can't see it from Earth.

For almost anything you want to see on the Moon, the best viewing time is when the object is near the *terminator,* which is the dividing line between bright and dark. Details of lunar features are most evident when features are just to the bright side of the terminator.

During a month, which is approximately the period of time from one full Moon to the next, the terminator moves systematically across the lunar near side so that, at one time or another, everything you can see on the Moon is close to the terminator. Depending on the time of the month, the terminator is either the place on the Moon where the Sun rises or the place where the Sun sets. As you know from experience on Earth, shadows extend farther during sunrise or sunset and continually shrink as the Sun gets higher in the sky. The length of the shadow when the Sun is at a known altitude is related to the height of the lunar feature that casts it. The longer the shadow, the taller the feature.

Sketching the Moon

You can view or photograph the Moon, and enjoy spotting maria, craters, and other topographic features marked on your lunar map. However, some amateur astronomers have a different hobby: They make drawings of lunar features seen through their telescopes. If you're artistically inclined, sketching lunar features may be an ideal astronomical hobby for you. The Moon is the only celestial object on which you can easily make out surface relief. You can do three-dimensional sketches of the lunar landscape, just as you may draw the view from your home. You can't do that for any other body unless you work from a space probe's close-up photo, not your own observations. *Sketching the Moon: An Astronomical Artist's Guide,* by Richard Handy, Deidre Kelleghan, Thomas McCague, Erika Rix, and Sally Russell, is a wonderful guide. It gives illustrated examples and step-by-step instructions for making your own telescopic drawings of the Moon by taking account of the interplay of light and shadow on the stark lunar surface.

Taking a shine to earthshine

When you observe the Moon, you may notice that the part on the dark side of the terminator isn't always pitch black. You may see a dim glow even though the Sun isn't shining there. The glow is earthshine, similar to the reddish glow on the lunar surface during a total eclipse of the Moon that I describe earlier in this chapter. *Earthshine* is sunlight that gets through Earth's atmosphere, where it is reddened (like the Sun at sunrise or sunset) and also bent slightly. It's bent enough to land on the Moon and cast a dim glow. You can see earthshine most easily when there's a crescent Moon, and you will never see it on a full Moon.

About the worst time to look at nearly anything on the Moon is during a full Moon. During a full Moon, the Sun is high in the sky on most of the lunar near side, so the shadows are few and short. The presence of shadows cast by features on the Moon helps you understand the *surface relief* — the way landforms extend above or below their surroundings.

Joining the dark side

You don't need a chart of the far side to help you observe the Moon because you can't see the far side. Our view is limited because the Moon is in *synchronous rotation,* meaning that it makes exactly one turn on its axis as it makes one orbit around Earth (the orbital period of the Moon is about 27 days, 7 hours, and 43 minutes). So the Moon always has the same hemisphere facing Earth.

Astronomy supply houses and science stores sell Moon globes that depict the features of the entire Moon, meaning the lunar near side and the far side. The Soviet space program first photographed the far side of the Moon, which it did by snapping pictures with a robotic spacecraft very early during the Space Age. Since then, many different spacecraft, including the Lunar Orbiters, Clementine, and the Lunar Reconnaissance Orbiter, have thoroughly mapped the Moon.

Going to lunar extremes: Bring your sunscreen, oxygen supply, and parka

When the Sun is up, the temperature on the lunar surface goes up to as much as 243°F (117°C), but at night, it drops to around −272°F (−169°C). These extreme temperature changes are due to the absence of any meaningful atmosphere to insulate the surface and reduce the amount of heat it loses at night. The Moon has no liquid water. The surface is too hot, too cold, and too dry to sustain life as we know it. And there's no air to breathe.

Being a lunar citizen scientist

You can study the Moon without a telescope and from the comfort of your home. Just surf to http://cosmoquest.org/mappers/moon/, click "Simply Craters" and follow simple directions. You'll learn to identify small craters on the Moon in photographs from NASA's Lunar Reconnaissance Orbiter.

Quite an impact: Considering a theory about the Moon's origin

Scientists know a lot about the ages of the rock in different terrains and parts of the Moon. They acquired the data with radioactive dating of samples from the hundreds of pounds of lunar rocks that the six crews of NASA Apollo astronauts — who landed on the Moon at different times from 1969 through 1972 — brought back to Earth.

Before the Apollo Moon missions, several top experts confidently predicted that the Moon would be the Rosetta stone of the solar system. With no liquid water to erode the surface, no atmosphere worth mentioning, and no active volcanism on the Moon, they thought the surface should include plenty of primordial material from the birth of the Moon and the planets. But the Apollo lunar samples threw rocks on their theory.

When a rock melts, cools, and crystallizes, all its radioactive clocks are reset. Radioactive isotopes begin producing fresh daughter isotopes that become trapped in the newly formed mineral crystals. The Apollo Moon rocks show that the whole Moon — or, at least, its crust down to a considerable depth — was melted well after 4.6 billion years ago. The very oldest surface rocks on the Moon are *only* 4.5 billion years old. The difference between 4.6 and 4.5 billion years is 100 million years. And unlike the minerals in Earth rocks, which contain water bound up into the mineral structures, the Moon rocks are almost bone dry.

The origin theory that has emerged to explain all this evidence, and to avoid the objections scientists posed against previous theories, is the *Giant Impact theory.* According to this theory, the Moon consists of material blasted out of the mantle of Earth by a huge object — with up to three times the mass of Mars — that struck young Earth a glancing blow. Some of the rock from the mantle of that long-vanished impacting object also was incorporated into the Moon, according to the theory.

The giant impact on young Earth knocked all this material up into space as a vapor of hot rock. It condensed and solidified like snowflakes. The snowflakes knocked into each other and stuck together, and before you knew it, the Moon had formed. It came together in powerful impacts of the last big pieces of accumulated rock, with the heat from each impact melting the rock.

All the impacts that caused the craters that we now see on the Moon happened later, and most of them date back to more than 3 billion years ago.

The Moon is less dense than Earth as a whole and about as dense as Earth's mantle (the layer beneath the crust and above the core), according to this theory, because it was made from mantle material. (Density is a measure of the amount of mass that's packed into a given volume. If you have two cannonballs of the same size and shape, they have the same volume. But if one ball is made of lead and one is made of wood, the lead ball is heavier and has a higher density.) This theory predicts that the Moon shouldn't have much of an iron core, if any. And a small core in a small object (meaning the Moon) should have cooled and frozen long ago if it ever contained liquid iron. Some lunar researchers suspect that, nevertheless, the Moon has an iron core and that it may be molten. NASA's twin-spacecraft GRAIL mission (Gravity Recovery and Interior Laboratory), launched in 2011, will eventually get to the bottom of this mystery.

The Giant Impact theory is currently our best guess. Unfortunately, we have no test for it at this time. For example, the theory predicts no special kind of rock that we can look for in the hundreds of pounds of lunar rocks that the Apollo astronauts collected. Some astronomers suspect that rocks in the biggest crater on the Moon were knocked out from a great depth by the colliding asteroid that formed the crater. Those rocks, in the South Pole–Aitken Basin, may come from beneath the surface layer that melted after the Moon formed. Studies of the rocks would then tell scientists whether the Giant Impact theory is accurate. The South Pole–Aitken Basin is the largest crater on the Moon — or anywhere else in the solar system — and is also the topic of big disagreements. Some experts think the rocks there don't come from deep enough to test the theory. Worse yet, plans to explore the Basin with astronauts or robotic craft are on hold as NASA struggles to devise a new Moon program.

If scientists verify the Giant Impact theory, it will have a giant impact on science, but don't hold your breath — this could take many moons.

I hope that you like my description of the Moon and that you won't be disappointed that it's "on the way out." The Moon is slowly moving away from Earth while still orbiting around it. Every year, the Moon recedes by about 1½ inches, and the time taken for one orbit around Earth gets very slightly longer.

Chapter 6

Earth's Near Neighbors: Mercury, Venus, and Mars

You can spot Earth's neighboring terrestrial (or rocky) planets Mercury, Venus, and Mars with the naked eye and inspect them with your telescope. But they tantalize you by revealing only a little of their nature in that way. Most of what scientists know about their physical properties, geologic forms, and histories is based on images and measurement data that interplanetary spacecraft sent back to Earth.

Two NASA spacecraft have visited Mercury, one flying past it three times and one orbiting the planet. Several probes have visited, orbited, and even landed on Venus. Mars has been the target of numerous probes, landers, and robot rovers. Some have orbited Mars, some have landed on it, and some, unfortunately, have crashed or even missed the planet.

In this chapter, I give you fascinating details about (and handy tips for viewing) Earth's closest neighbors in the solar system.

Mercury: Weird, Hot, and Mostly Metal

Astronomers have an ironclad case that Mercury is weird. It's not mostly rock, like Earth (or like the Moon, Mars, and Venus), but it's mostly metal — a big ball of iron with a thin skin of rock. On Earth, the iron core (see Chapter 5) extends just over halfway from the center to the surface. But on Mercury, the metal core extends 85 percent of the way to the surface, and it's all or nearly all molten. Mercury also has one layer that's unlike the layers inside any other planet: a zone of solid iron and sulfur between the iron core and the rocky surface.

Mercury's weirdness is plain to see on its surface as well. A great many impact craters pock the surface, as on the Moon (see Chapter 5), but many of Mercury's craters are tilted, as though the ground shifted after they formed. The largest crater on Mercury, the Caloris Basin (960 miles across) is strange, too: Much of the crater *bottom* is raised above the rim. Try to figure out how that happened!

Most of what we know about Mercury comes from NASA's Messenger probe, which is still orbiting the planet, and from Mariner 10, which flew by Venus in 1974 and visited Mercury three times in 1974 and 1975. You can view images from Messenger, get the latest news on Mercury discoveries, and keep up with the progress of the mission at its main website at The Johns Hopkins University (`http://messenger.jhuapl.edu`). You can also browse images from Mariner 10 at an Arizona State University site (`http://ser.sese.asu.edu/M10/IMAGE_ARCHIVE/Mercury_search.html`). Check out the color section of this book to see an image of Mercury.

Here are some other key facts about Mercury:

- Mercury has long, winding ridges that cut across impact craters and other geologic features. The ridges were probably caused by shrinkage of the crust, which was cooling from a molten state.

- Mercury has fewer small craters than the Moon, in proportion to the number of large craters.

- Highly cratered highlands are present on Mercury, as on the Moon (Mercury has no known moon of its own). But unlike on the Moon, Mercury's highlands are interrupted by gently rolling plains.

- On the antipode of the Caloris Basin, exactly opposite Caloris on the other side of Mercury, is a region of broken terrain. The collision that caused Caloris must have generated powerful seismic waves, which traveled through Mercury and around its surface, converging at the antipode with a catastrophic effect.

- Mercury has a global magnetic field, generated by a natural dynamo in its molten iron core, much like Earth, except that Earth's magnetic field is much stronger. (Mars, Venus, and the Moon all lack a global magnetic field.)

- Mercury has huge swings in temperature, from as much as 870°F during the day to as low as –300°F at night.

The chemistry of Mercury is strange, with lots of the volatile (easily evaporating) elements potassium, sodium, and sulfur on the surface. That finding challenges astronomers who are trying to explain how Mercury got to be mostly iron. Previous theories postulated that Mercury once had a larger proportion of rock than it does now. If so, it would have been more like Earth and the Moon. The theories proposed various ways of blasting away much of the original outer rock of Mercury, but any force so powerful would also have vaporized the potassium, sodium, and sulfur that now pervade the Mercury surface.

The more astronomers have learned about Mercury, the more it puzzles us.

Dry, Acidic, and Hilly: Steering Clear of Venus

Venus never sees a clear day; the planet is perpetually covered from equator to pole by a 9-mile-thick (15-kilometer) layer of clouds made up of concentrated sulfuric acid. And the surface has no relief from the heat: Venus is the hottest planet in the solar system, with a surface temperature of 870°F (465.5°C) that stays about the same from equator to pole, day and night.

And if the heat seems bad, check out the barometric pressure: It measures about 93 times the pressure at sea level on Earth. But forget about seas; you won't find any water on Venus. You can complain about the heat, but not the humidity — it's a dry heat, like in Arizona.

The bad news about the weather on Venus is that a perpetual rain of sulfuric acid falls all over the planet. The good news is that this rain is a *virga,* meaning rain that evaporates before it hits the ground.

The high temperature on Venus results from an extreme case of the *greenhouse effect.* In simple terms, the thick atmosphere (which is more than 95 percent carbon dioxide, with no oxygen) and clouds let through most of the sunlight that falls on Venus, which heats the surface and the air near the ground. The warm surface and air release heat in the form of infrared rays. On Earth, a lot of such infrared radiation escapes into space, cooling the ground at night. But on Venus, the carbon dioxide atmosphere traps the infrared, so the planet gets extremely hot.

Almost all the excellent images of the surface of Venus that you can find on NASA websites (and others) aren't photographs at all. What you see are detailed radar maps, notably from NASA's Magellan spacecraft. The clouds on the planet block the view of telescopes on Earth and of any camera on a Venus-orbiting satellite. The cloud tops are at an altitude of 40 miles (65 kilometers), much lower than where a satellite can operate.

The few images we have from Venus lander probes, launched by the former Soviet Union, show areas of flat rock plates separated by small amounts of soil. The plates resemble areas of hardened basalt lava flows on Earth. But on Venus, the surface appears orange because the thick cloud cover filters the sunlight. You can see satellite radar maps of Venus, lander images of the surface, and more at the Views of the Solar System website, at www.solar-views.com/eng/venus.htm. (You can see an image of Venus in the color section of this book, too.)

Flat plains that are volcanic lowlands with *rilles* (the winding canyons left by lava flows) cover the vast majority of Venus (about 85 percent). This territory includes the longest-known rille in the solar system, Baltis Vallis, which stretches across Venus for about 4,230 miles (6,800 kilometers). Cratered highlands and deformed plateaus are also present.

Not as many craters dot Venus as you may expect, based on the number you see on Earth's Moon (Venus has no known moon) and on Mercury. No small craters exist, and there aren't many large craters because Venus's surface was flooded with lava or reworked by volcanism (the eruption of molten rock from within a planet) after its bombardment by impacting objects had mostly ended. This flooding or reworking erased all or most of the early craters. Few large objects have struck Venus since the early craters were destroyed, and small objects don't make many craters on Venus: Aerodynamic forces in the thick Venus atmosphere impede and destroy objects capable of making craters up to 2 miles in diameter.

Huge volcanoes and mountain ranges cover the surface of Venus, but nothing resembling the nonvolcanic mountains on Earth (like the Rocky Mountains in the western United States or the Himalayas in Asia), which are caused by one crustal plate pushing into another. And Venus has no chains of volcanoes (like the Pacific "Ring of Fire"), which rise at the edges of plates. Plate tectonics and continental drift don't occur on Venus, as they do on Earth.

Venus looks different from Earth on the surface, but it bears an interesting resemblance underground. Venus Express, a probe sent by the European Space Agency, detected "hot spots," areas on Venus that are noticeably warmer than their surroundings. Astronomers believe that the hot spots are places where plumes of hot magma are welling up from deep below the surface. Earth has similar features, called mantle plumes, in regions where there are active volcanoes but that aren't near the borders of tectonic plates (for example, in Hawaii). (For more on tectonic plates, see Chapter 5.)

Red, Cold, and Barren: Uncovering the Mysteries of Mars

Scientists have mapped the surface of Mars and measured the altitudes of mountains, canyons, and other features with a high degree of accuracy. You can find the National Geographic chart of the entire planet at the NASA website (http://tharsis.gsfc.nasa.gov/ngs.html). The map is based on data from two instruments carried on the Mars Global Surveyor (MGS), a satellite that operated in orbit around Mars from 1997 to 2006. One instrument, a laser altimeter, bounced pulses of light off the Mars surface to measure the altitudes of the surface features that reflected the light. The other instrument, a camera, photographed landforms.

While MGS was still operating, another NASA spacecraft, the Mars Odyssey, arrived and began orbiting the planet in October 2001. You can see its findings at http://mars.jpl.nasa.gov/odyssey.

The European Space Agency doesn't get as much publicity as NASA, so you may not know that the Europeans have a Mars Express satellite that began orbiting the red planet on December 25, 2003. You can see splendid images from this spacecraft at www.esa.int/SPECIALS/Mars_Express.

Even though scientists have accurately mapped Mars with satellites and have explored parts of the Martian surface with lander probes and robot rovers, the planet holds many mysteries that they want to solve. In the following sections, I cover theories about water and life on Mars. (For even more on Mars, be sure to check out an image in the color section of this book.)

Where has all the water gone?

The topographic map of Mars shows that most of the Northern Hemisphere is much lower than the Southern Hemisphere. Some astronomers believe that this arid northern region, bounded by features that resemble an ancient shoreline, was an ocean long ago. Scientists used ground-penetrating radar mounted on Mars Express to peek under the possible shoreline region. In 2012, they reported that the subsurface layer resembles *sediment* (meaning broken rock, soil, and/or sand that settles to the bottom of a body of water). The sediment (if that's what it is) presumably was deposited in the suspected ancient sea. Some experts are still not convinced that Mars had an ocean, but I'm a believer.

The northern lowland may or may not be the site of an ancient ocean, but even if it isn't, other evidence suggests that liquid water was once common on Mars:

- Mars is cold and dry now, but there's a great deal of ice at the poles. By one estimate, enough ice is present to flood the entire planet to a depth of 100 feet if it melted. (The polar ice won't melt; Mars is just too cold.)

- Some canyons on Mars look like a great flood carved them out long ago.

- The Mars Reconnaissance Orbiter photographed many narrow linear features that run down steep slopes. They darken during the Martian summer and fade as the weather cools. They may be caused by salty water in the surface layer that melts and flows only when Mars is warm enough. (Saltwater stays liquid at lower temperatures than freshwater, so saltwater may flow on Mars when freshwater is solid ice.)

- The soil in many parts of Mars contains substances that form under the influence of water, including minerals that occur in clays.

- The planet has features that look just like dry riverbeds, with stream-lined islands and pebbles that seemingly were rounded in a torrent. Such pebbles were imaged by the Mars Pathfinder lander probe and its little robot, Sojourner.

- Mars Odyssey, taking instrument readings from orbit, found indications of large amounts of water, presumably frozen, just beneath the surface over large areas of Mars.

The Mars atmosphere is mostly carbon dioxide (as is Venus's atmosphere), but the Martian atmosphere is much thinner than the atmospheres of Earth or Venus. Mars also has clouds of water-ice crystals, which resemble the cirrus clouds on Earth. In winter, some carbon dioxide from the Martian atmosphere freezes on the surface, leaving thin deposits of dry ice. At the pole where winter is underway, a thin cap of dry ice often tops the permanent cap of water ice. If Mars had an ocean in the past, the planet had to have been much warmer than it is today. If the carbon dioxide atmosphere was much thicker then, it would have trapped heat, as in the greenhouse effect on Venus that I describe earlier in this chapter. If Mars once had a warm atmosphere and an ocean, then carbon dioxide from the atmosphere would have dissolved in the water. Chemical reactions then would have produced carbonates (minerals composed of carbon and oxygen). This theory predicts that there are carbonate rocks on Mars. NASA's Mars Exploration Rover, Spirit, discovered carbonate rocks in 2010!

Mars experts have a variety of opinions, but I think the case is closed: Mars once had a warm climate and a whole lot of liquid water.

Currently, it gets comfortably warm in the daytime on Mars at the equator, where noontime temperatures can reach a balmy 62°F (16.6°C). However, don't stay the night — it can get down to–208°F (–133.3°C) after sunset. The seasons on Mars differ from Earth's seasons, too. As I explain in Chapter 5, Earth's seasons are caused by the tilt of Earth's axis with respect to the plane of Earth's orbit around the Sun, _not_ by changes in Earth's distance from the Sun (which are negligible). On Mars, both the tilt of the planet's axis and the significant changes in its distance from the Sun from one place in Mars's orbit to another (because the orbit of Mars is more elliptical than Earth's almost-circular orbit) combine to produce "unearthly" seasons. Summer in the Southern Hemisphere on Mars is shorter and hotter than summer in the Northern Hemisphere, and winter in the Northern Hemisphere on Mars is shorter and warmer than winter in the Southern Hemisphere.

A magnetometer on MGS discovered long parallel stripes of oppositely directed magnetic fields frozen in the rocky crust of Mars. Mars doesn't have a global magnetic field today, but this finding may mean that it once had a global field that periodically reversed, just as Earth's field does (see Chapter 5). It may also mean that Mars endured a crustal process resembling the seafloor spreading on Earth and producing a similar pattern. But the molten iron core on Mars must have frozen solid long ago, so a new magnetic field is no longer generated, and the heat flow from the inside to the surface is so low that there's probably no volcanism still underway.

The volcanism that did occur on Mars produced immense volcanoes, such as Olympus Mons, which is about 370 miles (600 kilometers) wide and 15 miles (24 kilometers) high, or five times wider and almost three times higher than the largest volcano on Earth, Mauna Loa. Mars also has many canyons, including the immense Valles Marineris (Mariner Valley), which is 2,490 miles (4,000 kilometers) long. Impact craters dot the surface, too. The craters are more worn down than those on Earth's Moon because much more erosion has occurred, possibly caused by water that produced great floods on Mars (a controversial subject in astronomy to this day).

Does Mars support life?

People have many mistaken ideas about Mars, but some of the theories may actually be right; they just haven't been proven. These ideas all revolve around the possibility of life on Mars. Most of them are as improbable as the story about the future astronaut who returns from the planet: "Well, is there life on Mars?" the reporters demand. "Not much during the week," he says, "but on Saturday night. . . ."

Claims about life strike out

The discovery of the "canals" on Mars spawned the first widespread speculation about the possibility of life. Some of the most famous astronomers of the late 19th and early 20th centuries were among those who reported the canals. Planetary photography wasn't very useful in those days because the exposures were fairly long and atmospheric seeing (which I define in Chapter 3) blurred the images. So scientists believed that drawings by expert professional telescopic observers were the most accurate images of Mars. Some of these charts showed patterns of lines stretching and crisscrossing around the surface of Mars. Percival Lowell, an American astronomer, theorized that the straight lines were canals, engineered by an ancient civilization to conserve and transport water as Mars dried up. He concluded that the places where the lines crossed were oases.

Over the years, the idea of the "canals" and other reported indications of past or present life on Mars have struck out:

- ✔ When the American spacecraft Mariner 4 reached Mars in 1965, its photographs showed no canals, a conclusion verified in much greater detail by images from subsequent Mars probes. Strike one.

- ✔ Two later probes, the Viking Landers, conducted robotic chemical experiments on Mars to look for evidence of biologic processes such as photosynthesis or respiration. At first they appeared to have found evidence of biologic activity when water was added to a soil sample. But most scientists who reviewed the matter concluded that the water was reacting chemically with the soil in a natural process that doesn't involve the presence of life. Strike two.

- ✔ Viking Orbiters also sent back images of the Mars surface as they revolved around the planet. The images show, at one location, a crustal formation that — to some folks — looks like a face. Although many natural mountain peaks and stone formations on Earth resemble the profiles of the famous rulers, Native American tribal chiefs, and others for whom they're named, some true believers claim that the "face on Mars" is a monument of some type, erected by an advanced civilization. Later, sharper images from MGS showed that this landform doesn't look like a face at all. Strike three for the advocates of life on Mars.

But the idea of life was not "out," despite the three strikes. In 2003, astronomers began finding traces of methane (sometimes called "swamp gas") on Mars. Methane breaks down rapidly into other substances under Mars conditions, so some astronomers suggested that fresh methane is being generated by a primitive form of life on Mars. (On Earth, for example, microbes called methanogens give off methane.) However, geological processes on Mars also may make methane, so experts are still sniffing around this mystery.

The search for fossil evidence

In 1996, scientists analyzed samples of a meteorite that they believed came from Mars after being knocked off the planet by the impact of a small asteroid or comet. The scientists found chemical compounds and tiny mineral structures that they interpreted as chemical byproducts and possible fossil remains of ancient microscopic life. Their work is controversial, and many subsequent studies contradict these conclusions. Based on current research, scientists can't make a persuasive case that supports the theory of past life on Mars, nor can they disprove it.

The only action to take is to search systematically on Mars for evidence of life, past or present, in the regions that make the most sense — places where large quantities of water appear to have been present in the past and where layers of sediment were deposited in ancient lakes or seas. These types of places hold the most fossils on Earth.

In 2012, NASA launched the Mars Science Laboratory, carrying the Curiosity rover. The objective is to analyze Mars soil and rocks for evidence that past conditions were conducive to life. You can follow mission progress and learn about Curiosity's discoveries as they come in. Just head for the website at http://marsprogram.jpl.nasa.gov/msl/.

Differentiating Earth through Comparative Planetology

Mercury is a small world of extreme temperatures, with a global magnetic field like Earth's, but much weaker. Neither Venus nor Mars has such a magnetic field, although the two planets are similar to Earth in many other ways. But liquid water and life occur today only on Earth, as far as we know. What makes Earth different?

Venus, unlike Earth, has a hellish temperature. Venus is farther from the Sun than Mercury but is even hotter. The high temperature is due to an extreme *greenhouse effect,* the process by which atmospheric gases raise the temperature by absorbing outward flowing heat. Earth's atmosphere may once have contained large amounts of carbon dioxide, the way Venus's atmosphere does now. But on Earth, the oceans absorbed much of the carbon dioxide, so that gas couldn't trap as much heat in the atmosphere as it does on Venus.

Mars, on the other hand, is too cold to support life (as far as we can tell). Mars has lost most of its original atmosphere, and its current atmosphere isn't thick enough to produce a greenhouse effect sufficient to warm much of the surface above the freezing point of water often or for very long.

The three large terrestrial planets are like the bowls of cereal in the child's story of Goldilocks. Venus is too hot, Mars is too cold, but Earth is *just right* to support liquid water and life as we know it. (Little Mercury is also too hot.) Putting together the information on the basic properties of the terrestrial planets and their respective differences, scientists can conclude that

- ✔ Mercury is like the Moon on the outside (with lots of craters) but like Earth on the inside, with a molten iron core that generates a magnetic field.

- ✔ Venus is Earth's "evil twin," It's about the same size as Earth, but with deadly heat and pressure, an unbreathable atmosphere, and highly acid rain.

- ✔ Mars is the little Earth that cooled and dried up.

Earth is the Goldilocks planet — just right!

When you contrast the properties of the planets like this, you can draw conclusions about their respective histories and why those different histories have brought the planets to their present states. Think that way, and you're practicing what astronomers call *comparative planetology*.

Observing the Terrestrial Planets with Ease

You can spot Mercury, Venus, and Mars in the night sky with the aid of monthly viewing tips from astronomy magazines and their websites, a smartphone app, or a desktop planetarium program (see Chapter 2 for information on all these resources). Venus is especially easy to find because it's the brightest celestial object in the night sky other than the Moon.

Mercury is the planet that orbits closest to the Sun, and Venus is next. They both orbit inside the orbit of Earth, so Mercury and Venus are always in the same region of the sky as the Sun, as seen from Earth. Therefore, you can find these planets in the western sky after sunset or in the eastern sky before dawn. At those times, the Sun isn't far below the horizon, so you can see objects near and to the west of the Sun in the morning before sunrise, and you can see objects near but east of the Sun in the evening after sunset. Your motto as a Mercury or Venus spotter is "Look east, young woman" or "Look west, young man," depending on whether you skywatch at dawn or dusk and whether you're a fan of old Western movies.

A bright planet appearing in the east before dawn is commonly called a *morning star,* and a bright planet in the west after sunset is an *evening star.* As Mercury and Venus move swiftly around the Sun, this week's morning star may be the same object as next month's evening star (see Figure 6-1).

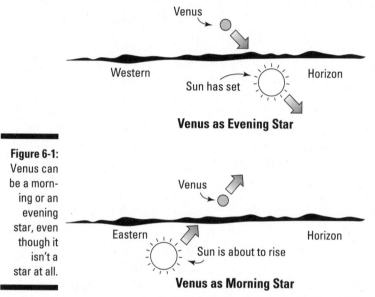

Figure 6-1: Venus can be a morning or an evening star, even though it isn't a star at all.

Dinah L. Moché/Astronomy: A Self-Teaching Guide, Seventh Edition

In the following sections, I explain the best time to observe the terrestrial planets based on elongation, opposition, and conjunction — three terms that describe the planets' positions in relation to the Sun and Earth — and how to use this understanding in your observations of the terrestrial planets. (I list the planets in the order of ease of observation, starting with Venus, the easiest.)

Understanding elongation, opposition, and conjunction

Elongation, opposition, and *conjunction* are terms that describe a planet's position in relation to the Sun and Earth. You encounter these terms when you check listings of the planets' positions to plan your observations. Here's what the terms mean:

- ✔ *Elongation* is the angular separation between a planet and the Sun, as visible from Earth. Mercury's orbit is so small that the planet never gets more than 28° from the Sun. During some periods, it doesn't get farther than 18° from the Sun, making it hard to spot. Venus can get up to 47° from the Sun.

   Greatest western (or eastern) elongation occurs when a planet is as far from the Sun as it can get during a given *apparition* (a period of time when the planet is visible from Earth on successive nights). Some greatest elongations are greater than others because sometimes Earth is closer to the planet than at other occasions. Elongation is especially important when you observe Mercury because the planet is usually so close to the Sun that the sky at its position isn't very dark.

- ✔ *Opposition* occurs when a planet is on the opposite side of Earth from the Sun. Opposition never happens for Mercury or Venus, but Mars is at opposition about once every 26 months. This is the best time to observe the planet because it appears largest in a telescope. And at opposition, Mars is highest in the sky at midnight, so you can view it all night.

- ✔ *Conjunction* occurs when two solar system objects are near each other in the sky, such as when the Moon passes near Venus as we see them. In reality, Venus is far beyond the Moon, but we see a conjunction of the Moon and Venus.

Conjunction has a technical meaning as well. Instead of describing positions in *right ascension* (the position of a star measured in the east–west direction) and *declination* (the position of a star measured in the north–south direction), astronomers sometimes use ecliptic latitude and longitude. The ecliptic is a circle in the sky that represents the path of the Sun through the constellations. *Ecliptic latitude* and *longitude* measure degrees north and south (latitude) or east and west (longitude) with respect to the ecliptic. (Don't worry; you don't need to use the ecliptic system when you observe the terrestrial planets. But knowing about it helps you understand the definitions of *superior* and *inferior conjunction* that follow.)

You need to master some tricky terminology to understand conjunctions and oppositions — namely, the labeling of planets as superior and inferior and the labeling of conjunctions as superior and inferior. A *superior planet* orbits outside the orbit of Earth (so Mars is a superior planet, for example). An *inferior planet* orbits inside the orbit of Earth (so Mercury and Venus are inferior planets — in fact, they're the only inferior planets).

When a superior planet is at the same longitude as the Sun as seen from Earth, it's on the far side of the Sun and is said to be at *conjunction* (see Figure 6-2). When that same planet is on the opposite side of Earth from the Sun (also shown in Figure 6-2), it's at *opposition.*

Conjunction is a bad time to observe a superior planet because it's on the far side of the Sun and also in nearly the same direction. So don't try to observe Mars at conjunction; you won't see it. The best time to observe Mars is at opposition.

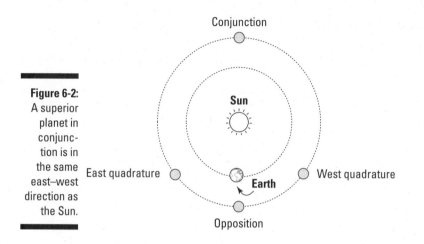

Figure 6-2:
A superior
planet in
conjunc-
tion is in
the same
east–west
direction as
the Sun.

A superior planet has conjunctions and oppositions, but an inferior planet
has two kinds of conjunction and never has an opposition (see the diagram in
Figure 6-3). When the inferior planet is at the same longitude as the Sun and
is between the Sun and Earth, it's at inferior conjunction. And when the inferior planet is at the same longitude as the Sun but is beyond the Sun as seen
from Earth, the planet is at superior conjunction.

If you can explain all that to your friends, you'll truly feel superior! Feel free
to do the explaining "in conjunction" with Figures 6-2 and 6-3.

You can see Venus best at inferior conjunction, when it looks largest and
brightest, but Mercury is too close to the Sun as seen from Earth for viewing
at inferior or superior conjunction. The best time to see Mercury is when it is
at greatest elongation.

Figure 6-3:
Inferior
conjunc-
tion aligns
an inferior
planet
with the
Sun in the
east–west
direction.

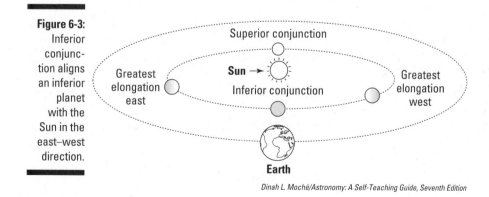

Dinah L. Moché/Astronomy: A Self-Teaching Guide, Seventh Edition

Viewing Venus and its phases

The simplest planet to find is Venus. The second rock from the Sun is so bright that people with no knowledge of astronomy frequently notice it and call radio stations, newspapers, and planetariums to ask what "that bright star" is.

When scattered clouds move from west to east in front of Venus, inexperienced viewers often misunderstand the scene. Folks think that Venus (which they don't recognize) is moving rapidly in the opposite direction of the clouds. Due to its brightness and the mistaken impression that it's moving rapidly behind a cloud deck, people often report Venus as an unidentified flying object. It isn't. Astronomers know it well.

When you become familiar with Venus, you may be able to spot it in broad daylight. Often Venus is bright enough that, if the sky is clear, you can spot it in the daytime by using *averted vision.* In other words, you can glimpse it "out of the corner of your eye." For some reason, you may be able to spot a celestial object easier with averted vision than if you look right at it. The ability to spot things with averted vision may be a survival trait; it makes it a little harder for an enemy or a predator to sneak up on you from the side.

Wait just an arc minute (or second)

Scientists measure apparent sizes in the sky in angular units. Something that goes all the way around the sky, such as the celestial equator, is 360° long. The Sun and the Moon, in comparison, are each about a half-degree across. The planets are much smaller, so smaller units are necessary to describe them. A degree is divided into 60 minutes of arc, and a minute of arc — called an *arc minute* or *arc min* — is divided into 60 seconds of arc — usually called an *arc second* or *arc sec.* A degree is made up of 3,600 (60 × 60) seconds of arc. In many astronomy books and articles, a single prime symbol (') represents a minute of arc, and a double prime (") represents a second of arc. Readers often mistake these symbols as abbreviations for feet and inches. You can tell when a clueless copy editor has had the last cut at an astronomy article because you see a statement like "The Moon is about 30 feet in diameter."

Venus is actually only about 5 percent smaller in diameter than Earth. Its apparent size, or *angular diameter,* ranges from about 10 arc sec when Venus is farthest away (and with a full Moon shape) to about 58 arc sec in diameter when it's closest (and is a narrow crescent).

Here's an interesting factoid for you: An arc second equals the apparent size of a dime (U.S. ten-cent coin) seen from a distance of about 2 miles.

A small telescope can show you Venus's most recognizable features: its phases and changes in apparent size. Venus has phases similar to Earth's Moon (see Chapter 5), and for the same reason: Sometimes part of the Venus hemisphere that faces the Sun (and is, therefore, bright) is directed away from Earth, so a telescopic view of Venus shows a partly illuminated and partly dark disk.

The dividing line between the bright and dark parts of Venus is called the *terminator,* just as on the Moon. Don't worry: This terminator is just a perfectly safe imaginary line on Venus (see Chapter 5).

As Venus and Earth orbit the Sun, the distance between the two planets shifts substantially. At its closest to Earth, Venus is a mere 25 million miles away; at its farthest, it's a full 160 million miles distant. What's important here is the proportional change: At closest approach, Venus is about six times nearer to Earth than at its most distant position. Therefore, it looks six times bigger through a telescope.

What you don't see when you view Venus are striking features, like the Man in the Moon. Venus is totally covered by thick clouds, and all you can see is the top of the clouds. Venus is so bright because it orbits relatively close to both the Sun and Earth and because it has a nice, bright reflecting cloud layer. But sometimes you may be able to discern the horns of the Venus crescent extending farther into the dark side than predicted for the phase on that day. In this case, you see some sunlight that has bounced around in Venus's atmosphere and passed beyond the terminator into the side of the planet where night has fallen.

Images of Venus with striking cloud patterns, like those you see in books, were made in ultraviolet light where the patterns show up. Ultraviolet light doesn't pass through our atmosphere (hurray for the ozone layer, which blocks this hazardous radiation), so you can't view Venus in it. In fact, you can't see ultraviolet light anyway; it's invisible to the human eye. But telescopes on satellites and space probes above or beyond the atmosphere can take ultraviolet pictures.

On rare occasions, observers report a pale glow on the dark part of Venus. This glow, called the *ashen light,* is sometimes real and sometimes a trick of the imagination. After centuries of study, experts still can't explain the ashen light, so some of them even deny that it exists. But with luck, you may see it. People claim to see other features on Venus through their telescopes, but almost all the reports are wrong. Experiments show that the reports are usually due to a psychological effect: If people view a featureless white globe from a distance, they may discern patterns that don't exist.

Try to catch a planetary transit

On rare occasions, you can see a *planetary transit,* when a planet passes right in front of the Sun and looks like a tiny black disk against the bright solar surface. Only the inferior planets (Mercury and Venus) transit the Sun because only they come between Earth and the Sun.

You can observe a transit of Mercury with a small telescope, but you must use the procedures for safe solar viewing that I explain in Chapter 10. Otherwise, you risk incurring severe eye damage and even blindness. It may be a good idea to check out whether your local planetarium, museum, or astronomy club is holding a transit-viewing event (they often do). Then you can look with equipment set up by experts. The next two transits of Mercury are on May 9, 2016, and November 11, 2019.

Transits of Venus are more impressive than those of Mercury because Venus has a bigger disk, but Venus transits are also much rarer. Unfortunately, if you haven't already seen a Venus transit, I think you never will. The last one was in June 2012, shortly before this book went to press, and the next transit of Venus isn't until December 2117. Leave a note for your descendants to be sure and watch.

Watching Mars as it loops around

Mars is a bright red object, but it isn't nearly as dazzling as Venus. So check your sky maps to make sure that you don't mistake a bright red star, such as Antares in Scorpius (whose name means "rival of Mars"), for the red planet.

The great advantage in observing Mars is that, when it appears in the night sky, it often remains visible for much of the night, unlike Mercury and Venus, which set fairly soon after sunset or rise only shortly before dawn. You usually have time for dinner and the nightly news before you head into the backyard to check on Mars.

With a small telescope, you can spot at least a few dark markings on Mars. The best periods to see the features last a few months but occur only about every 26 months, when Mars is at opposition. At opposition, Mars looks biggest and brightest, and you can see details more easily on its surface.

The upcoming oppositions of Mars are in

April 2014

May 2016

July 2018

October 2020

December 2022

January 2025

Don't miss them!

At its best oppositions — when it looks biggest and brightest — Mars is south of the celestial equator, but you can still observe it from temperate latitudes in the Northern Hemisphere.

The easiest Martian surface feature to spot with a small telescope is usually Syrtis Major, a large dark area extending northward from the equator. Mars's day is nearly the same as Earth's: 24 hours, 37 minutes. So if you look at Mars off and on during a night, you may be able to see Syrtis Major move slowly across the planet's disk as Mars turns. Experienced amateur planetary observers may see its polar caps and other markings as well.

NASA images of Mars, taken by interplanetary probes and the Hubble Space Telescope, are much too detailed to guide you in small-telescope observation. You need a simple *albedo map,* which charts and names the bright and dark areas on Mars as visible with small telescopes. An albedo map offers more detail than the average observer ever sees, and it offers a good guide and a challenge to your observing skills. You can find such a map (prepared by the Association of Lunar and Planetary Observers) on the NASA MarsWATCH website, at `http://mars.jpl.nasa.gov/MPF/mpf/marswatch/marsnom.html`. I also recommend *A Traveler's Guide to Mars,* a quality paperback with foldout charts, by William K. Hartmann, a leading planetary scientist who is a fine space artist, to boot (Workman Publishing, 2003).

Astronomers rate sky conditions in terms of *seeing* (the steadiness of the atmosphere above the telescope), *transparency* (the freedom from clouds and haze), and *sky darkness* (the freedom from interfering artificial light, moonlight, or sunlight). When observing a bright planet such as Mars, good seeing is the most important factor, and dark sky is least important. But the darker the sky, the steadier the air, and the higher the transparency, the more you can enjoy the night.

With good seeing, the stars don't twinkle quite as much, and you can use a higher-magnification eyepiece with the telescope to bring out fine details on Mars or another planet. When the seeing isn't good, the telescopic image looks blurred and seems to jump around. Under adverse conditions, high magnification is useless; you only magnify the blurred, jumping image. Use a low-power eyepiece for the best results.

Tracking Mars's backtracking

A basic project for beginning planet gazers is to track the motion of Mars across the constellations; all you need are your eyes and a sky map.

Locate Mars among the stars and mark that position with soft pencil on your map. If you repeat this observation on each clear night, you can see a pattern emerge that puzzled the ancient Greeks and led to complicated theories — most of them wrong.

Most of the time, Mars moves eastward from night to night, just as Earth's Moon moves eastward across the constellations. The Moon keeps going, but Mars sometimes reverses course. For two to almost three months (62 to 81 days) at a time, Mars heads west across the constellations, moving backward by 10° to 20°. After this period, it gets back on track and heads east again. The backtracking is called Mars's *retrograde motion.*

The backtracking isn't a case of Mars not knowing whether to come or go. The retrograde motion is just an effect produced by Earth racing around the Sun. While you chart Mars's motion, you stand on Earth, which races around the Sun once every 365 days. Mars moves slower, making one full orbit in 687 days. As a result, when we pass Mars on our inside track (lapping it), Mars seems to move backward against the reference frame of the distant stars. But in reality, Mars always forges ahead.

Unfortunately, even when atmospheric conditions are ideal at your observing site and when an opposition of Mars is in progress, disaster may strike. Mars is a planet that experiences worldwide dust storms, which hide its surface features from view.

In fact, professional astronomers sometimes rely on amateur astronomers to help monitor Mars, to let them know when a dust storm begins and to report other pronounced changes in the appearance of the planet. You can get information on this program at the International MarsWatch website, at Rowan University (http://elvis.rowan.edu/marswatch/news.php). On the MarsView page at this site, you can plan your observation of Mars by inserting the date and UTC time when you plan to observe (I explain how to get the UTC time in Chapter 5, in the section, "Finding the right time"). Next, the website displays a color picture that predicts what Mars will look like at your specified time. If you see a dust storm covering much of Mars, it may spoil your view, but you could be credited for alerting scientists to the event. Experts will welcome your dust story, not brush it off.

You need experience to become a reliable telescopic Mars observer. As a beginning observer, don't assume that a great dust storm is in progress just because you can't make out any detail. Get accustomed to seeing Mars on different occasions, when seeing conditions are sometimes better and sometimes worse. Only then should you consider that, when you can't see details, the planet is at fault, not the viewing circumstances or your inexperience. Remember this scientific motto: "The absence of evidence is not necessarily evidence of absence." You may not see Mars surface details the first time you look, but that doesn't mean that a dust storm is obscuring your view. As a telescopic observer, you have to train your viewing skills, just as gourmets and wine lovers train their palates. I'll drink to that.

And just for your information, Mars has only two known moons, Phobos and Deimos. These tiny celestial bodies aren't visible with small telescopes.

Outdoing Copernicus by observing Mercury

Historians say that the great Polish astronomer Nicholas Copernicus (1473–1543), who proposed the *heliocentric* (Sun-centered) *theory* of the solar system, never spotted the planet Mercury.

But Copernicus didn't have modern aids, such as smartphone apps, desktop planetariums, astronomy websites, or even monthly astronomy magazines (see Chapter 2). You can use these aids to find out when Mercury will be best placed for observation during the year: the times of greatest western and eastern elongation (terms that I cover in the section "Understanding elongation, opposition, and conjunction," earlier in this chapter), which occur about six times each year.

At temperate latitudes, such as those of the continental United States, Mercury is usually visible only in twilight. By the time the sky is dark, well after sunset, Mercury has set, too. And in the morning, you can't spot Mercury until the impending dawn starts to light the sky. Mercury resembles a bright star but appears much dimmer than Venus in the west at dusk or in the east before dawn.

You can get more information on observing Mercury and the other planets from the Association of Lunar and Planetary Observers (ALPO), which also collects sketches and other planetary observations made by amateur observers and offers observing forms, charts, and other publications. ALPO has separate sections (subgroups) for observers of Mercury and other planets. Check out its website at http://alpo-astronomy.org. In the United Kingdom, the British Astronomical Association has similar groups; go to its website at http://britastro.org/baa and click "Sections" for more information.

Be an early riser for Mercury

Mercury is much smaller than Venus, but you can see its phases through your telescope. The best time to do this is when Mercury is at western elongation and appears in morning twilight. The atmospheric steadiness, or seeing, is almost always better low in the east near dawn than it is low in the west after sunset, so you get a much sharper view in the morning. Standard guides, like the highly regarded *Observer's Handbook* (an annual publication of The Royal Astronomical Society of Canada; www.rasc.ca), and the astronomy magazines and their websites (see Chapter 2) all tell when the elongations are due. If you want the information long in advance, consult the useful handbook *Celestial Delights: The Best Astronomical Events Through 2020,* by Francis Reddy (Springer, 2012) for a table of the greatest elongations of Mercury and Venus through the year 2020.

You need a viewing site with a clear eastern horizon because Mercury doesn't get high in the sky when the Sun is below the horizon. If you have trouble spotting it with your naked eye, sweep around that part of the sky with a pair of low-power binoculars. And if you have a computerized telescope with a built-in database, you can just punch in "Mercury" and let the telescope do the finding.

Don't expect to see surface markings

Seeing surface markings on Mercury with a small telescope, or with almost any telescope on Earth, is extremely difficult. Mercury's apparent size at greatest elongation is only about 6 to 8 arc sec (see the sidebar "Wait just an arc minute [or second]," earlier in this chapter, for details).

Some experienced amateur observers report seeing surface markings, but no useful information has ever come from such sightings. A few of the greatest planetary observers of all time thought that they could see and draw the surface markings. From their drawings, the observers tried to deduce the rotation period, or "day," of Mercury. They concluded that Mercury's day was equal to the length of the year on Mercury, or 88 Earth days. But they were wrong. Radar measurements later proved that Mercury turns once every 59 Earth days. So a year on Mercury is less than two of its days.

In any case, when you find out how to spot Mercury by eye and then view its phases with your telescope, you'll be way ahead of Copernicus!

Mercury lovers choose morning

Here's why seeing is better near the dawn horizon than near the sunset horizon: By sunset, the Sun has warmed Earth's surface all day, so as you look out low in the sky to the west, you look through turbulent currents of warm air rising from the surface. But in the morning, Earth has had all night to cool down and stabilize. It takes a few hours for the rising Sun to warm the land and mess up the seeing again.

Chapter 7

Rock On: The Asteroid Belt and Near-Earth Objects

*A*steroids are big rocks that circle the Sun. The vast majority of asteroids are safely beyond the orbit of Mars in an area called the *asteroid belt,* but thousands of other asteroids follow orbits that come close to or cross Earth's orbit. Many scientists believe that an asteroid hit Earth about 65 million years ago, wiping out the dinosaurs and many other species.

In this chapter, I introduce you to these big rocks and explain the best ways to observe them. And in case you're worried, I tell you the truth about the risk of an asteroid hitting Earth in the future and fill you in on the research scientists are conducting to deal with the possibility.

Taking a Brief Tour of the Asteroid Belt

Asteroids are also called minor planets because when they were first discovered, experts thought they were objects like the planets. But astronomers now believe that asteroids are remnants of the formation of the solar system — objects that never merged with enough additional space debris to grow into planets. Some asteroids, such as Ida, even have their own moons (see Figure 7-1). Asteroids are made of silicate rock, like the rocks of Earth, and of metal (mostly iron and nickel). Some asteroids may also contain carbonaceous (or carbon-bearing) rock. And in recent years, ice has been found on some asteroids.

Figure 7-1:
The asteroid
Ida has its
own moon,
Dactyl.

Most of the known asteroids are in a huge, flat region centered on the Sun and located between the orbits of Mars and Jupiter. We call this region the *asteroid belt.* Asteroids range in size from Ceres, which is 592 miles (952 kilometers) in diameter, to large meteoroids, which are just fragments of asteroids (see Chapter 4). A boulder-size space rock is either a very small asteroid or a very large meteoroid; feel free to take your pick.

Asteroidal meteoroids, which I describe in Chapter 4, are made of rock and/ or iron; when they fall to Earth, they are called meteorites. You can see those meteorites in many natural history and geology museums. Asteroids are just much bigger hunks of the same stuff. To check out some meteorites, consult the Meteor Collector website, at www.meteoritecollector.org/museums. html, and find a museum collection near you.

Table 7-1 lists the four biggest objects in the asteroid belt. The two largest, Ceres and Pallas, are at nearly the same average distance from the Sun, although Pallas has a much more elliptical orbit.

Table 7-1	The "Big Four" of the Asteroid Belt		
Name	*Diameter in Miles*	*Diameter in Kilometers*	*Mean Distance from the Sun (AU)*
Ceres	592	952	2.77
Pallas	339	545	2.77
Vesta	329	530	2.36
Hygiea	253	407	3.14

As of late 2011, there were about 600,000 known asteroids, of which roughly 16,000 had been named. (That count includes one that the International Astronomical Union was kind enough to name after me; I'm glad they didn't just call it "Dummy.") Most were found in recent years by robotic telescopes designed for the purpose, but experienced amateur astronomers who mount advanced digital cameras on their telescopes are also making discoveries.

You can readily see the largest asteroids, such as Ceres and Vesta, through small telescopes (see the later section "Searching for Small Points of Light" for more about observing asteroids).

Ceres and Vesta are so big that their own gravity makes them round. One crater on Vesta, Rheasilvia, is almost as wide as the asteroid itself. Deposits of dark material on Vesta's surface may be from the impacts of carbon-bearing meteoroids or from collisions with small asteroids that melted the surface and caused lava flows. There's probably a lot of ice in cold, dark places on Vesta, too.

Small asteroids are often shaped like potatoes and frequently look blasted apart (see Figure 7-2) because, indeed, they have been. The asteroids in the belt constantly collide with each other, making impact craters and breaking off big and little chips. (It gives new meaning to the slang verb "bump off.") The big chips are simply smaller asteroids, and the little chips are asteroidal meteoroids.

At rare intervals, small asteroids (or large meteoroids) smash into Earth (see the next section for more on this phenomenon). Asteroid impacts (and comet impacts) have also covered the Moon, Mars, and Mercury with craters; Venus has craters, too, but not as many. (I describe these cratered objects in Chapters 5 and 6.)

Asteroids have craters as well, but they're much harder to see with telescopes because the asteroids themselves are so small. In most telescopes, an asteroid is just a point of light, like a star. You can glimpse craters and other features on Vesta at the website of NASA's Dawn space probe. Dawn arrived at Vesta in July 2011 to study it in detail before departing for Ceres, where it will arrive in 2015. To see the pictures and learn about the probe's discoveries as Dawn sheds early light on these asteroids, visit http://dawn.jpl.nasa.gov.

An earlier NASA probe paved the way for Dawn by exploring the 21-mile-long asteroid Eros, which crosses inside the orbit of Mars. The Near Earth Asteroid Rendezvous–Shoemaker probe orbited Eros for a year and then landed on February 12, 2001. You can see a video of this oblong asteroid rotating at the NEAR Shoemaker website (www.discovery.nasa.gov/near.cfml).

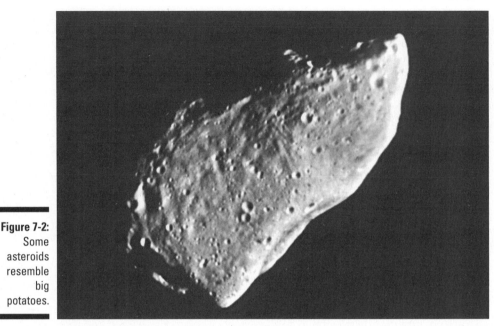

Figure 7-2:
Some
asteroids
resemble
big
potatoes.

Courtesy of NASA

Understanding the Threat That Near-Earth Objects Pose

Not all asteroids are orbiting safely beyond Mars. Thousands of small asteroids follow orbits that cross or come close to Earth's orbit. Astronomers call these neighbors *near-Earth objects (NEOs),* and they've classified 1,296 of them as *potentially hazardous asteroids (PHAs),* as of March 2012. Some day, one of these scary neighbors may come uncomfortably close to Earth or even strike our planet. The Minor Planet Center (MPC) of the International Astronomical Union keeps tabs on PHAs, and several observatories sweep the skies to discover more of them.

The MPC's website (www.minorplanetcenter.net) offers a mix of information for experts and amateur astronomers alike, including maps of the inner and outer solar system, updated daily, that show where the planets and many of the asteroids are located in space.

Astronomers don't know of any specific object that's currently a menace to Earth. But a rock a few miles in size that strikes Earth at 25,000 miles per hour (11 kilometers per second) would cause a far greater catastrophe than the simultaneous explosion of all the nuclear weapons ever made. It would be a rare case when astronomy isn't fun. Asteroids about that big collide with Earth every 10 million years, give or take, and smaller ones hit more often. Larger asteroids impact Earth much less frequently because, the bigger the asteroids, the fewer there are of them.

Conspiracy theorists think that if astronomers did know about a doomsday asteroid, we wouldn't tell. But face it, if I knew the world was in danger, I'd settle my affairs and head for the South Seas instead of sitting around finishing this chapter!

In 1998, the Hollywood movies *Armageddon* and *Deep Impact* gave sensationalized versions of what can happen if a large asteroid or comet was on a collision course for Earth. Such catastrophe stories are inspired, in part, by the widely accepted conclusion that an asteroid about 6 miles (10 kilometers) wide struck Earth about 65 million years ago. The Chicxulub crater, a 110-mile-wide (180-kilometer-wide) geologic formation resting partly on Mexico's Yucatan peninsula and partly offshore in the Gulf of Mexico, may be the surviving trace of the impact, which is theorized to have wiped out the dinosaurs.

The action of weather and geological processes, such as mountain building, erosion, and volcanism, has eroded the impact craters on Earth and eliminated most of them. You can see the list of more than 180 of our planet's remaining craters, with photos and maps of some of them, at the Earth Impact Database website of the University of New Brunswick. Just surf to www.passc.net/EarthImpactDatabase/index.html.

Flooding, erosion, volcanism, and other processes disturb the landscape of Earth and bury, fill in, or partly destroy many impact craters — and some of them haven't been found yet. If you do a lot of hiking, exploring, or aerial sightseeing, you may even discover an impact crater. If you spot a possible crater, consult the Earth Impact Database website, mentioned earlier, and you'll find an online form to report your discovery.

An impact by a small asteroid caused the famous Meteor Crater (which ought to be called Meteoroid Crater or Asteroid Crater) in northern Arizona, near Flagstaff. The site is well worth a visit because it's the largest well-preserved impact crater on Earth. When I was a young scientist at the Kitt Peak National Observatory in Arizona during the 1960s, you could just tell the receptionist at Meteor Crater that you were an astronomer and get in free. Nowadays, everybody over the age of 5 is charged admission, but it's worth it. Check out the website at www.meteorcrater.com for information on the hours the crater is open to visitors, directions, and more.

For a brief time in March 1998, many people feared that a small, newly discovered NEO could strike Earth in the year 2028. Astronomers eliminated that possibility within a day when additional observations showed that the asteroid's orbit won't intersect Earth. Some experts even disagreed with the initial prediction — as experts often do.

Although Earth appears to be safe for now, scientists may discover an NEO on a collision course with Earth in the future, so they spend time now studying the options for what can be done in such an eventuality.

If you want to explore the consequences of a collision, download the CraterSizeXL app for your iPhone or iPad. Insert the size and speed of a falling body, and the app calculates the size of the crater that it will make. Or you can visit the Impact Earth! website at Purdue University (www.purdue.edu/impactearth) and calculate the catastrophe online.

When push comes to shove: Nudging an asteroid

Some experts propose developing a powerful nuclear missile to intercept a killer asteroid before it can strike Earth. But if we blow up an asteroid heading our way, the results could be worse than the damage caused by the impact of the intact asteroid. It would be like the scene in the Walt Disney movie *Fantasia* in which the sorcerer's apprentice chops up the out-of-control magic broom that won't stop fetching water. He ends up with a whole bunch of little brooms, each of which starts fetching water.

If we blow up an asteroid with a nuclear bomb, a swarm of smaller rocks would follow the same deadly trajectory. The rocks would pack more wallop than all the Pentagon's weapons combined. A better idea is to use the nuclear missile (or perhaps some other kind of missile) to only nudge the asteroid so that it passes through its intersection point with Earth's orbit a little early or a little late, when Earth hasn't reached that spot yet or has already gone past. Phew!

The problem with nudging an asteroid is that scientists don't know how much force to apply. We don't want to break it up, but because we don't know the mechanical strength of asteroids, we don't know how hard to hit it. Asteroids may be made of strong rock or fragile stone. Some may be mostly solid metal. If we don't know our enemy, we can make things worse by striking in the wrong way.

Rather than risk unwanted consequences by trying to blow up or poke a threatening asteroid, engineers suggest using a *gravity tractor.* A massive spacecraft would fly along with the asteroid for a period of years. Without ever touching the threatening body, the spacecraft would slowly change the rate at which the asteroid is moving, thanks to the gravitational attraction between the two bodies. This plan should keep the asteroid intact but moving so that it just misses Earth. The problem with this method is that it involves launching a very heavy spacecraft near the asteroid and then keeping it close to the big rock for a decade or more. Not enough time may remain to do this after the collision is predicted.

Experts have proffered a lot of other ideas about how to avoid a threatening asteroid, but it's not clear which method is best.

In *Fantasia,* the sorcerer himself breaks the spell on the enchanted broom, but without a sorcerer to make the asteroids disappear, we need hard information to design a system that can safely protect Earth from asteroids.

Forewarned is forearmed: Surveying NEOs to protect Earth

Astronomers have a plan to help design a system that can protect Earth from renegade asteroids:

1. **Take a census of the NEOs to make sure that we've located every rock that measures a kilometer or more in size in our region of the solar system.**

   NEOs of this size can become PHAs if their orbits take them close to Earth.

2. **Track these NEOs and compute their orbits to determine whether any are likely to strike Earth in the foreseeable future.**

3. **Study the physical properties of asteroids to discover as much as we can about them.**

   For example, make telescopic observations to determine what kind of rock or metal they are made of.

4. **When astronomers understand the threat, an engineering team can design a space mission to counteract it.**

To survey the NEOs, special-purpose asteroid-discovery telescopes are in operation at several locations. You can visit their websites and see their recent findings. Two of the most important projects are listed here:

- ✔ The Lincoln Near Earth Asteroid Research (LINEAR) project, at White Sands Missile Range, New Mexico, supported by the U.S. Air Force and NASA (`www.ll.mit.edu/mission/space/linear`)
- ✔ NASA's Near-Earth Asteroid Tracking (NEAT) project, observing from the Maui Space Surveillance Site in Hawaii and Palomar Observatory in California (`http://neat.jpl.nasa.gov`)

Astronomers are finding NEOs and determining their orbits. Experts then can calculate the odds that a given asteroid will strike Earth in the near future. But — and this may surprise you — nobody is in charge of reacting to an asteroid collision threat if one is predicted. Defense departments and military commands worldwide are responsible for protecting their national territories and (sometimes) the territories of allied nations. But no space agency or armed force is charged with the mission of defending Earth against a threat from space. Let's hope that a protective agency is established and given the necessary power and resources by the time we all need it. Otherwise, we will truly be between a rock and a hard place.

Searching for Small Points of Light

Looking for asteroids is like scanning the sky for comets (see Chapter 4), except that you look for a small point of light resembling a star rather than a fuzzy image. But unlike a star, an asteroid moves perceptibly against the background of other stars from hour to hour and from night to night.

You can easily see the largest asteroids, such as Ceres and Vesta, through small telescopes; astronomy magazines publish short articles and sky charts to guide you in advance of good viewing periods (in general, there are no best times of the day or year to view asteroids). Most good desktop planetarium programs (and similar smartphone apps) display sky maps that point out the location of asteroids. (See Chapter 2 for more about magazines, apps, and planetarium programs; go to Chapter 3 to find out about telescopes.)

You aren't ready to search systematically for unknown or "new" asteroids until you become a skilled amateur astronomer with a few years of experience. Advanced amateurs search for new asteroids with electronic cameras on their telescopes. They collect a series of images of selected areas of the sky, generally in the direction opposite the Sun (which, of course, is below the horizon). When they see a small point of light (resembling a star) change its location, they probably see an asteroid.

The easiest asteroid-related activity for beginners to try is observing occultations. An *occultation* is a kind of eclipse that occurs when a moving body in the solar system passes in front of a star. The bodies responsible can be Earth's Moon (lunar occultation), the moons of other planets (planetary satellite occultation), asteroids (asteroidal occultation), or planets (planetary occultation). Comets and the rings of planets can also cause occultations. An occultation doesn't look like much; you just see the star disappear for a short time during the eclipse.

You can enjoy an asteroidal occultation without obtaining scientific data, but what a waste of a unique opportunity! The details of an occultation differ from place to place on Earth. For example, the same occultation may last longer as seen from one place on Earth than from another, or it may not occur at a certain location. So at some locations, you see the star being eclipsed, and at other locations, you see the star without an eclipse. From occultation data, astronomers can get a more accurate picture of a number of sky objects. For example, sometimes the occultations reveal that what seems to be an ordinary star is actually a close *binary system* (two stars in orbit around a common center of mass; see Chapter 11 for details on binary stars).

The following sections tell you how to track and time asteroidal occultations.

Helping to track an occultation

Asteroidal occultations are much trickier to observe than lunar occultations because astronomers often can't predict them with sufficient precision. Astronomers go to various spots on the predicted *occultation ground track* (a narrow band across the surface of Earth where astronomers expect the occultation to be visible — just like the path of totality in an eclipse of the Sun, which I describe in Chapters 2 and 5) and attempt to observe asteroidal occultations. But because the diameters, orbits, and shapes of most asteroids aren't known with sufficient accuracy, the predictions can't be precise. Because the occultation may be visible at some places and not others, astronomers need volunteers to monitor an asteroidal occultation at many locations. Amateur observations help determine the sizes and shapes of the asteroids involved in the occultations. You can join in, too.

Report your observations to the International Occultation and Timing Association (IOTA); check out its website at www.occultations.org. On the site, you can download a free IOTA Observers Manual and fill in and submit an Asteroidal Occultation Report Form. The IOTA website is regularly updated to provide the latest predictions of occultations by asteroids and other objects, so be sure to check it periodically.

IOTA recommends that you begin occultation study by observing with an experienced astronomer, just to get the hang of it. I bet you'll enjoy every iota of it!

Timing an asteroidal occultation

To make your asteroidal observation scientifically useful, you need to time it accurately and know the exact location (latitude, longitude, and altitude) of where you are when you observe the occultation. In the past, observers figured out their locations by consulting topographic maps. Nowadays, you can use a GPS receiver or smartphone app to determine the coordinates of your observing site.

Chapter 8

Great Balls of Gas: Jupiter and Saturn

- -

In This Chapter

▶ Understanding the recipe for gas giant planets

▶ Spotting the Great Red Spot and Jupiter's moons

▶ Observing Saturn's rings and moons

- -

*J*upiter and Saturn, located beyond Mars and the asteroid belt, are among the best sights to see through a small telescope, and at least one is usually well placed for observations. The four largest moons of Jupiter and the famous rings of Saturn are the favorite targets when amateur astronomers give friends and family some peeks through their telescopes. And although you may not be able to tell through the telescope, the underlying science of these huge planets and their satellites is fascinating, too. In this chapter, I describe the magnificent sights you can observe through your telescope and clue you in on the basic facts about the two largest planets in our solar system.

The Pressure's On: Journeying Inside Jupiter and Saturn

Jupiter and Saturn are like hot dogs with unapproved food coloring. The meat isn't the mystery; the additives are. What you see in telescopic photographs of Jupiter and Saturn are the clouds, made of ammonia ice, water ice (like the cirrus clouds on Earth), and a compound called ammonium hydrosulfide. Water-drop clouds may be a part of the mix, too. But appearances are deceiving. These cloud materials are made of trace substances. Jupiter and Saturn are mostly hydrogen and helium, like the Sun. And despite much theorizing, scientists have no proof of what makes the Great Red Spot on Jupiter red or what produces the other off-white tints in the clouds of the two great planets.

Jupiter and Saturn are the largest of the four gas giant planets (the others are Uranus and Neptune). Jupiter has 318 times the mass of Earth; Saturn surpasses Earth's mass by about 95 times. As a result, their gravity is enormous, and inside the planets, the weight of the overlying layers produces enormous pressure. Descending into Jupiter or Saturn is like sinking in the deep sea. The farther down you go, the higher the pressure. And unlike the sea, the temperature increases radically with the depth. Don't even think about scuba diving there.

Up at the atmospheric levels where astronomers can see, in the cloud decks, the temperatures drop to –236°F (–149°C) on Jupiter and –288°F (–178°C) on Saturn. But at great depths, the squeeze is on. By the time you reach 6,200 miles (10,000 kilometers) below the clouds on Jupiter, the pressure soars to 1 million times the barometric pressure at sea level on Earth. And the temperature equals that of the visible surface of the Sun! But Jupiter is weirder than the Sun. The density of the thick gas at this depth is much higher than at the solar surface, and the hot hydrogen gas is compressed so that it behaves like a liquid metal. Swirling currents of this liquid metal hydrogen generate powerful magnetic fields on Jupiter and Saturn that reach far out into space.

Jupiter and Saturn glow intensely in infrared light, each generating almost as much energy as it gets from the Sun. (Earth, on the other hand, derives almost all its energy from the Sun.) The upward-moving heat, together with heat from the downward-shining rays of the Sun, stirs up their atmospheres and produces jet streams, hurricanes, and other kinds of atmospheric storms that continually change the appearance of these planets.

Almost a Star: Gazing at Jupiter

Jupiter has about 1,000th of the mass of the Sun. Sometimes scientists call it "the star that failed." If only it had 80 or 90 times more mass, the temperature and pressure at its center would be so high that nuclear fusion would begin and keep going. Jupiter would start shining with its own light, making it a star!

Jupiter has a diameter of about 88,700 miles (143,000 kilometers), which is about 11 times bigger than Earth. The gas giant rotates at enormous speed, making one complete turn in only 9 hours, 55 minutes, and 30 seconds. In fact, Jupiter turns so fast that the rotation makes it bulge at the equator and flatten at the poles. With a clear look in steady air, you can detect this *oblate* shape through your telescope.

The rapid spin helps produce ever-changing bands of clouds, parallel to Jupiter's equator. What you see through your telescope when you view Jupiter is really the top of the planet's clouds. Depending on the viewing conditions, the size and quality of your telescope, and circumstances on Jupiter itself, you may see from as few as 1 to as many as 20 cloud bands (see Figure 8-1).

Figure 8-1:
Jupiter and
its spinning-
induced
bands of
clouds.

Courtesy of NASA

Jupiter's darker bands of clouds are called belts; the lighter bands are zones. When you look through a telescope, Jupiter looks like a round disk. Right down the center of the disk is the Equatorial Zone, flanked by the North and South Equatorial Belts (NEB and SEB). In the SEB, you may see the Great Red Spot, often the most conspicuous feature on Jupiter. This atmospheric disturbance, sometimes compared to a great hurricane, has hovered in the Jupiter atmosphere for at least 120 years. In fact, the Great Red Spot may have been spotted as early as 1664, although if so, it faded away before reappearing in the 19th century.

Jupiter is easy to find because, like Venus (see Chapter 6), it shines brighter than any star in the sky. (A small exception: When its orbit takes it to the far side of the Sun, Jupiter may be slightly fainter than the brightest star, Sirius.) If you have a computer-controlled telescope that can point to the position of the planet, sometimes you can see Jupiter in the daytime. In exceptional circumstances, you may locate Jupiter with binoculars or even with the naked eye in the daytime. A deep blue sky with little or no airborne dust helps — as do smartphone apps such as Google Sky Map or Sky Safari (which I describe in Chapter 2).

When you can spot Jupiter with ease, you're ready for slightly more detailed observations. I provide directions for spotting the planet's features and moons in the following sections.

Scanning for the Great Red Spot

The Great Red Spot, shown in Figure 8-2, is a storm as big as Earth and some-times bigger in the South Equatorial Belt. Like most of Jupiter's features, it can change from day to day. Its color can grow paler or deeper. White clouds, which are big enough to see with some amateur telescopes, form near the spot and move along the SEB. Sometimes a cloud in the SEB or another belt seems to be drawn out across the planet, stretched mostly in longitude. A cloud with this linear shape is known as a *festoon,* and spotting this interest-ing display is indeed a festive occasion!

If you don't see the Great Red Spot at first, it may be in a pale condition, but more likely, it has rotated to the backside of Jupiter. You have to wait for Jupiter to rotate back around. If you take telescopic views of the features of Jupiter at intervals of an hour or two during the night, you can see the spot and smaller features move across the disk of the planet as Jupiter rotates.

In the early 1990s, the SEB itself seemed to disappear overnight. Later it reappeared. This belt has faded away and then appeared again several times since then. Amateur astronomers are sometimes the first to spot SEB events like these. So while you enjoy the spectacle of Jupiter's belts and spots, be alert for something new.

Figure 8-2:
Jupiter's
Great Red
Spot makes
for stormy
viewing.

Courtesy of NASA

Jupiter's invisible accessories

Did you know? Jupiter has thin rings made up of rock particles, perhaps tiny fragments from numerous boulders in the rings. Unlike Saturn's famous rings, Jupiter's accessories are dark, making them invisible through amateur telescopes. In fact, the rings are hard to see through almost any telescope, except the Hubble, and are best observed with instruments that space probes carry right up to Jupiter.

Microscopic particles from the rings are lifted into a thick halo interior to the rings, which surrounds and perhaps merges into the upper atmosphere of Jupiter.

NASA's Galileo probe, which operated around Jupiter from 1995 to 2003, obtained terrific images of the rings. You can find pictures of Jupiter's rings and more at NASA's Planetary Photojournal site, `http://photojournal.jpl.nasa.gov/index.html`. Just click on the image of Jupiter and follow the links.

Shooting for Galileo's moons

Whenever the seeing is good, your telescope will reveal structure in Jupiter's cloud tops — belts, zones, spots, and maybe more — and you may see one or more of the planet's four large moons: Io, Europa, Ganymede, and Callisto. (Check out a photo of Jupiter and these moons in the color section of this book.)

Jupiter's four prominent moons (it has 62 smaller known moons as of March 2012) are known as the Galilean moons, or Galilean satellites, named after Galileo, their discoverer. Each of the big four moons orbits almost exactly in the equatorial plane of Jupiter, so each moon is always right overhead somewhere on Jupiter's equator. Any telescope worth owning can spot the Galilean moons, and many people can even see two or three of them through a good pair of binoculars. However, Io, the innermost of the Galilean moons, is hard to spot through binoculars because it orbits very close to the bright planet.

You can't see enough detail on any of Jupiter's moons with your own telescope to figure out what their surfaces are like, but you can notice differences in their brightnesses and, with careful study, perhaps in their colors.

If you take a look at space probe images of the Galilean moons, you can see that each moon is a little world unto itself, with composition and landscape that give it individual character. (See the section "Timing your moon gaze," later in this chapter, for websites where you can check out the images.)

For basic details on the four major moons, check out the following list:

✔ **Callisto:** Callisto has a dark surface marked by many white craters. The surface is probably dirty ice — a mixture of ice and rock. The impacts of asteroids, comets, and big meteoroids have exposed the underlying clean ice — hence, the white craters.

✔ **Europa:** This moon has a ridged terrain that looks like rafts of ice. The surface may be a frozen crust that tops off an ocean of water and slush, perhaps 90 miles (150 kilometers) deep. Europa is one of five places in the solar system outside of Earth where scientists have strong evidence that liquid water is present underground. (The others are Ganymede, Callisto, and Saturn's moons Titan and Enceladus.)

✔ **Ganymede:** At 3,270 miles in diameter (5,262 kilometers), Ganymede is the largest moon in the solar system (even larger than 3,032-mile-wide Mercury). Ganymede's blotchy surface consists of light and dark terrains, perhaps ice and rock, respectively. The most noticeable marking is Valhalla, a huge-ringed impact basin about as large as the continental United States (judging the size by the outermost ridge).

✔ **Io:** This moon's surface is peppered with more than 400 volcanoes. Io is the only place other than Earth where we have definite evidence of ongoing volcanism as we know it on Earth, with hot lava emerging from underground. (But see my description of Saturn's moon Enceladus, later in this chapter, for a case of icy volcanism.) Io has no visible impact crater because lava from the ubiquitous volcanoes has covered up all impact sites.

Although you can't enjoy the kind of up-close and personal view achieved with sophisticated space equipment, you can observe with your telescope some interesting aspects of these moons as they orbit Jupiter. I cover phenomena that may affect your view of the moons — such as occultations, transits, and eclipses — in the following sections.

Recognizing moon movements

Io, Ganymede, Europa, and Callisto are always moving, changing their relative positions and appearing and disappearing as they revolve around Jupiter. Sometimes you can see them all, and sometimes you can't. If you can't spot one of the moons, here are a few possible explanations:

✔ An *occultation* may be underway, which occurs when one of the moons passes behind the limb of Jupiter (the edge of the disk you see through your telescope).

✔ The moon may be in *eclipse,* which occurs when the moon moves into Jupiter's shadow. Because Earth is often well to the side of a straight line from the Sun to Jupiter, Jupiter's shadow can extend well to its side as you see it from Earth. When you see a moon in plain view off the limb of Jupiter and it suddenly dims and disappears, it has moved into the planet's shadow.

✔ The moon may be *in transit* across the disk of Jupiter; at that time, the moon is particularly tough to see because the moons are pale in color, making them hard to spot against the cloudy atmosphere of Jupiter. In fact, a moon in transit can be much more difficult to discern than its shadow.

You can also observe a *moon shadow,* which occurs when one of the moons is sunward of Jupiter and casts a shadow on the planet. The shadow is a black spot, much darker than any cloud feature, moving across the planet. The moon that casts the spot may be in transit at the time, but this isn't always the case. When Earth is well off the Sun–Jupiter line, you may see a moon off the limb of Jupiter casting a shadow on the planet.

Timing your moon gaze

Sky & Telescope magazine carries a monthly schedule of the occultations, eclipses, shadow events, and transits of the four Galilean moons. Both that publication and *Astronomy* also print a monthly chart showing the positions of the four moons with respect to the disk of Jupiter night by night. (See Chapter 2 for more about astronomy magazines.) You can tell which moon is which by comparing what you see through the telescope with the chart.

Remember the following general rules as you watch Jupiter's moons:

✔ All four Galilean moons orbit around Jupiter in the same direction. When you see them on the near side, with respect to Earth, they move from east to west; when they orbit on the far side, they travel from west to east.

✔ A transiting moon is moving westward, and a moon about to be occulted or eclipsed is heading eastward (following the east–west geographic directions in the sky of Earth).

Under excellent viewing conditions, you can discern one or two markings on Ganymede, the largest Galilean moon, with a 6-inch or larger telescope. (See Chapter 3 for information on telescopes.) But to see the details of the surface, you need an image from an interplanetary spacecraft that visited the Jupiter system.

The best images of Jupiter and its moons are from the Galileo and Voyager 1 and 2 space probes and from the Hubble Space Telescope:

✔ You can find the Galileo images at `http://solarsystem.nasa.gov/galileo/gallery/index.cfm`. On that web page, just click the picture of Jupiter or one of its moons to bring up the images of that object.

✔ The Voyager images, plus some others, are at NASA's Planetary Photojournal website, `http://photojournal.jpl.nasa.gov`. When you see the planets, click the picture of Jupiter.

✔ For Hubble images, examine the collection at the Space Telescope Science Institute, at `http://hubblesite.org/newscenter/archive/releases/solar-system/jupiter`. The newest Jupiter probe, Juno, is en route to the planet, with a scheduled arrival in August 2016. You can follow its progress at the website `www.nasa.gov/mission_pages/juno/main/index.html`.

Com(et)ing within striking distance

On rare occasions, a comet or an asteroid strikes Jupiter, which may cause a temporary dark blotch atop the cloud belts that can last for months. Scientists didn't know this until July 1994, when large chunks of the broken comet Shoemaker-Levy 9 struck Jupiter. Since then, astronomers have gone back through old records of the markings on Jupiter and have found some suspicious features that may have been created in the same way.

Since 1994, astronomers have known that if a dark blotch occurs on Jupiter, it may be debris from an impacting object, not just another cloud formation on the belted planet. In July 2009, amateur astronomer Anthony Wesley discovered a fresh blotch on Jupiter with his 14.5-inch telescope near Canberra, Australia. He alerted scientists, and the Hubble Space Telescope photographed the blotch, which measured 5,000 miles across. (The impacting body was much smaller, just as a burning house can be dwarfed by the cloud of smoke that it produces.) In June 2010, Wesley observed a brief flash of light near the edge of Jupiter's disk. It was the glow from a large meteoroid fall on Jupiter, and another amateur in the Philippines recorded it on video. Keep your eyes (and telescope!) peeled for unusual features as you observe Jupiter.

If you spot something new and exciting on Jupiter or Saturn (which I describe next), send a report to the International Outer Planet Watch (http://atmos.nmsu.edu/ijw/ijw.html), the Jupiter Section of the Association of Lunar and Planetary Observers (www.alpo-astronomy.org/jupiterblog), or the British Astronomical Association's Jupiter Section (www.britastro.org/jupiter). Check these websites for useful observing information and recent reports on Jupiter.

Our Main Planetary Attraction: Setting Your Sights on Saturn

Saturn is the second-largest planet in our solar system, with a diameter of about 75,000 miles (121,000 kilometers). Most people are familiar with Saturn because of its striking set of rings. For centuries, astronomers thought that Saturn was the only planet that had rings. Today we know that rings encircle all four giant gas planets: Jupiter, Saturn, Uranus, and Neptune. But most of the rings are too dim to see through telescopes from the ground. The great exception is Saturn.

According to many observers, Saturn is the most beautiful planet. Not only are its famous rings easily visible through almost any telescope, but you can also spot Saturn's giant moon, Titan. Although many astronomers find Saturn's rings to be the celestial sight that most impresses their nonastronomer friends, Titan is also a worthy attraction.

In the following sections, I provide information on observing Saturn's rings, storms, and moons. Be sure to check out images of Saturn in the color section, too.

Ringing around the planet

Saturn's rings are usually easy to see because they're large and composed of bright particles of ice — millions of little ice fragments, some larger ice balls, and maybe some pieces the size of boulders. You can enjoy the rings through a small telescope and make out their shadows on the disk of Saturn (see Figure 8-3). Under excellent viewing conditions, the *Cassini division* — a gap in the rings named for the person who first reported it — may also be discernable.

Measuring more than 124,000 miles (200,000 kilometers) across, Saturn's rings are only yards (or meters) thick. Proportionately, the rings are like "a sheet of tissue paper spread across a football field," as Professor Joseph Burns of Cornell University once wrote. But even though the rings are proportionately as thin as facial tissue, you wouldn't want to blow your nose in them. Stuffing ice up your nostrils may chill you out more than sniffing glue, but I definitely don't recommend it.

Saturn spins once every 10 hours, 39 minutes, and 22 seconds and is even more oblate — flattened at the poles — than Jupiter. The rings tend to mislead the eye a little, however, so noticing Saturn's squashed shape can be tricky.

Figure 8-3:
Fragments of ice and rock make up Saturn's rings.

Courtesy of NASA

The rings are very large but also very thin. They keep a fixed orientation, pointing face-on at one direction in space. There's a time each year when the rings are more face-on than usual, as seen from Earth, and a time three months later when they come closer to edge-on than usual.

As Saturn goes around its own 30-year orbit, sometimes the rings are precisely edge-on and seem to vanish through small (or sometimes even large) telescopes. You can't see the rings when their edges face Earth because they're extremely thin. On those occasions, with a powerful telescope, you may see the rings projected as a dark line against the disk of Saturn. The rings last disappeared in 2009 and will vanish again in 2025.

Storm chasing across Saturn

Saturn has belts and zones just like Jupiter (see the section "Almost a Star: Gazing at Jupiter," earlier in this chapter), but Saturn's have less contrast and are much harder to see. Look for them during times of good atmospheric conditions, when you can use a higher-power eyepiece on your telescope to spot planetary details.

About once every 20 to 30 years, a big white cloud, or "great white storm," appears in the Northern Hemisphere of Saturn. High-speed winds spread out the cloud until it forms a thick, bright band all the way around the planet. After a few months, it may disappear. Sometimes amateur astronomers are the first to spot a new storm on Saturn. The last great white storm began in 2010, so you may have to wait a good while to see another. In the meantime, keep an eye out for smaller white clouds that can grow and spread partway around the planet.

Monitoring a moon of major proportions

Titan, Saturn's largest moon, is bigger than the planet Mercury. Its diameter is 3,200 miles (5,150 kilometers). Some large moons have thin atmospheres, but Titan has a thick, hazy atmosphere, composed of nitrogen and trace gases such as methane. Titan's atmosphere is hard to see through, but in 2004, the NASA Cassini space probe started mapping Titan's surface in infrared light (good for penetrating haze) and radar (even better). On January 14, 2005, the European Space Agency's Huygens probe landed on Titan. Most of what we know about this unusual moon comes from Cassini and Huygens.

Much of Titan's surface is relatively flat and smooth. At higher latitudes, the moon has lakes of liquid hydrocarbons, probably methane and ethane. (Hydrocarbons are various chemical compounds composed of hydrogen and carbon atoms; on Earth, they occur naturally in the crude oil that comes out of the ground.) On Earth, we have the *water cycle,* in which water rains down to the ground; flows into rivers, lakes, and seas; evaporates as water vapor; and rises into the atmosphere, from which it rains down again. Titan has a similar cycle, but it involves methane rain, methane lakes, and gaseous methane. Titan is so cold that any water is permanently frozen. Some hydrocarbons also freeze: In dry spots on Titan, there are "sand" dunes, but they're not made of rock particles, like the sand dunes on Earth. Titan's dunes most likely consist of frozen hydrocarbon particles. (You've probably got some solid hydrocarbons in your home, such as polystyrene, the plastic used in disposable cups for hot coffee or tea.)

If there were water on the surface of Titan, it would all be frozen. In 2012, Cassini discovered that there is a water ocean about 60 miles below the surface on Titan, where conditions are warmer.

Making sense of a cryptic moon

One of Cassini's most interesting discoveries was the existence of vents in the south polar region of Saturn's satellite Enceladus. Water vapor, ice particles, and other substances are pouring out of the vents, like cold versions of the geysers of hot steam in Yellowstone National Park. The eruption of ice-cold material like this is called *cryptovolcanism.* Astronomers concluded that the Enceladus geysers are fed from a subsurface body of liquid water on the moon that's warm enough to support life, if anything lives there. Fresh ice crystals from the geysers coat the surface of Enceladus, making it especially bright. Some ice particles are propelled out into space, where they merge into one of Saturn's rings.

With a good small telescope, you can see Titan. You may be able to see two more of Saturn's moons, Rhea and Dione, when they're near their largest elongations from the planet. (See Chapter 6 for more on elongations.) You can find a monthly chart of the locations of these moons with respect to the disk of Saturn in *Sky & Telescope* magazine. Use the charts to plan your observations. As of March 2012, Saturn had 62 known moons, most of them too small to see through amateur telescopes.

The Cassini probe is in the midst of a long tour of Saturn, and its moons that should extend until at least 2017. You can see pictures and other data that Cassini sends back at the Cassini Solstice Mission page, `http://saturn.jpl.nasa.gov`, and at the Cassini Imaging Central Laboratory for Operations (CICLOPS) site, `www.ciclops.org`.

Born moons and converts orbiting in harmony

Moons come in two varieties: regular and other. The regular moons all orbit in the equatorial plane of their planets, and they all orbit in the same direction in which the planets spin on their axes. This direction is called *prograde*. The regular moons almost certainly formed in place around Jupiter and Saturn, from an equatorial disk of protoplanetary and protomoon material. So Jupiter and Saturn, together with their many moons, are like miniature solar systems centered on big planets rather than stars.

But some of the small moons are like Elsa, the lioness that was "born free" and taken captive as a cub. They orbit in the direction opposite to the way the planet spins. These orbits are called *retrograde,* and they often are tilted with respect to the equatorial planes of their planets. The retrograde-orbiting moons formed elsewhere in the solar system, perhaps as asteroids, and were captured by the gravity of Jupiter and Saturn.

Jupiter has 66 confirmed moons, and Saturn has 62, as of March 2012. Each planet probably has quite a few more small ones, and astronomers keep finding more of them. Any number of known moons that you find in a printed book may be obsolete by the time you read it. Sometimes astronomers announce moons but don't count them. The officials at the International Astronomical Union want to be sure that the discoveries are confirmed. You can check the latest information on the natural satellites of Jupiter, Saturn, and other planets at the NASA Solar System Dynamics site, `http://ssd.jpl.nasa.gov/?sat_discovery`. The moons without names are provisional discoveries awaiting confirmation.

Chapter 9

Far Out! Uranus, Neptune, Pluto, and Beyond

Although Mars and Venus are closer to Earth, and Jupiter and Saturn are the bright, showy planets, observing the outer planets has its own mystique and offers its own rewards. This chapter introduces you to our solar system's two outermost planets — Uranus and Neptune — and describes Pluto (now called a dwarf planet). I also provide details on the moons of Uranus, Neptune, and Pluto; offer useful tips for viewing these far-out worlds; and share details about the Kuiper Belt.

Breaking the Ice with Uranus and Neptune

The following are the most important facts about Uranus (pronounced *yoo-RAN-us,* or more commonly *YOO-rin-us*) and Neptune:

✔ They have a similar size, with similar chemical compositions.

✔ They're smaller and denser than Jupiter and Saturn.

✔ Each planet is the center of a miniature system of moons and rings.

✔ Each planet shows signs of a long-ago encounter with a large body.

The atmospheres of Uranus and Neptune, like those of Jupiter and Saturn (see Chapter 8), are mostly hydrogen and helium. But astronomers call Uranus and Neptune *icy planets* because their atmospheres surround cores of rock and water. The water is so deep inside Uranus and Neptune and is under such high pressure that it's a hot liquid. But when the planets merged and coalesced from smaller bodies billions of years ago, the water was frozen.

You can tell a bona fide planetary scientist from a layperson because the scientist calls the hot water inside Uranus and Neptune "ice," whereas the civilian innocently calls the hot water "hot water." Scientists use technical jargon the way predatory animals use scent markings — to proclaim their exclusive territory.

Uranus has about 14.5 times the mass of Earth, and Neptune equals 17.2 Earths, but they appear nearly the same size. The lighter Uranus is a bit larger, measuring 31,770 miles (51,118 kilometers) across the equator. Neptune's equatorial diameter is 30,775 miles (49,528 kilometers).

One day on Uranus lasts about 17 hours and 14 minutes; a day on Neptune spans 16 hours and 7 minutes. So as with Jupiter and Saturn, these planets both rotate faster than Earth. Although the days on Uranus and Neptune are shorter than on Earth, the years are much longer. Uranus takes about 84 Earth years to make one trip around the Sun, and Neptune takes about 165 Earth years.

I cover more interesting facts about each planet in the following sections. Be sure to check out photos of Uranus and Neptune in the color section.

Bull's-eye! Tilted Uranus and its features

The evidence that Uranus suffered a major collision or gravitational encounter is that the planet seems to have flipped on its side. Instead of the equator being roughly parallel to the plane of Uranus's orbit around the Sun, it nearly forms a right angle with that plane so that, in terms of Earth's directions, its equator runs almost north–south.

Sometimes the North Pole of Uranus points toward the Sun and Earth, and sometimes the South Pole faces our way. For about a quarter of Uranus's 84-year orbit around the Sun, its North Pole faces roughly sunward; for about another quarter, the South Pole faces roughly sunward; and the rest of the time, the Sun illuminates the whole range of latitudes from pole to pole. In 2007, the Sun was overhead at the equator on Uranus. That period would have been a good time to go to the beach, if Uranus had a beach. On Earth, the Sun is never high in the sky at the North or South Pole, but in 2028, the Sun will be high over the North Pole on Uranus.

Observations with the Hubble Space Telescope and earlier, with the Voyager 2 space probe, show that Uranus has a changing pattern of cloud belts. In 2006, a large dark spot appeared. The changing cloud patterns on Uranus may be related to the seasons on that planet.

As of March 2012, Uranus has 27 known moons. It also has a set of rings. The rings are made of very dark material, probably carbon-rich rock, like certain meteorites known as carbonaceous chondrites. The moons and rings of Uranus orbit in the plane of its equator, just as the Galilean moons orbit in the equatorial plane of Jupiter (see Chapter 8), so the rings and the orbits of the moons of Uranus are at nearly right angles to the plane of the planet's orbit around the Sun.

Essentially, you can think of Uranus and its satellites as a big bull's-eye that sometimes faces Earth and sometimes doesn't. One or more large objects probably struck Uranus long ago and tilted it from its natural position.

Against the grain: Neptune and its biggest moon

Neptune isn't tilted from the natural order; its equator is parallel to the plane of its orbit around the Sun, or nearly so. Its rings are very dark, like those of Uranus, and probably consist of carbon-bearing rock.

Neptune has 13 known moons as of March 2012. Its largest moon, Triton (which is larger than Pluto), has a diameter of 1,682 miles (2,707 kilometers). Seen from north and above, Neptune, like all the planets in our solar system, revolves counterclockwise around the Sun. Most moons revolve counterclockwise around their planets. But Triton, which resembles a cantaloupe in photos from Voyager 2, goes against the grain, traveling clockwise around Neptune. (In other words, it has a retrograde orbit, a term that I define in Chapter 8.) After mulling it over, scientists concluded that Neptune captured Triton early in solar system history. Expert opinions vary, but a leading theory is that, back then, Neptune had a near collision with a binary system of two small objects from the Kuiper Belt (which I discuss at the end of this chapter). Neptune grabbed Triton, which was one half of the binary, and the other little planet flew away. Astronomers need more facts to be able to test this theory.

Triton consists of ice and rock, so it seems more like Pluto (see the next section) than Uranus or Neptune. Its surface is shaped by eruptions and flows of cold icy substances rather than hot, molten rock. (It's the result of *cryovolcanism,* which I describe in Chapter 8.) Water ice, dry ice, frozen methane, frozen carbon monoxide, and even frozen nitrogen are all present on Triton. The moon doesn't have many impact craters, probably because they got sloshed full of ice over time.

Environmental groups say that excessive tourism endangers national parks, so consider a trip to Triton instead. Its landscape is just as bizarre, and maybe as beautiful, as Yellowstone's. But if you head for Triton, expect a "Winter Wonderland." The surface has cold surges rather than hot springs, and its geysers spew long plumes of frigid vapor rather than torrid jets of steam. Just bring a space suit and some warm booties.

Neptune's atmosphere features cloud belts, and occasionally a so-called Great Dark Spot appears, which may be a very large storm, similar to the Great Red Spot on Jupiter (see Chapter 8). Jupiter's big spot appears, fades, and reappears again, all in about the same place in the same cloud belt. But Neptune's Great Dark Spot was first found in the planet's Southern Hemisphere in 1989, then it disappeared, and later a so-called Great Northern Dark Spot came into view in the opposite hemisphere.

Meeting Pluto, Planet or Not

For decades, astronomers deemed Pluto the most distant planet from the Sun in our solar system (see Figure 9-1). Actually, Pluto moves inside the orbit of Neptune every 248 years for a few decades at a time, but the last such inside move ended in early 1999. It won't happen again in the lifetime of anyone now inhabiting Earth, unless medical research makes major strides long before the 23rd century. But Pluto was demoted. On August 24, 2006, the International Astronomical Union (IAU) voted to demote it to the status of *dwarf planet.*

Dwarf planets are a newly recognized class of astronomical objects, defined by the IAU as bodies with these characteristics:

- ✔ Orbit the Sun directly (as opposed to orbiting another body, such as a planet)
- ✔ Are massive enough that their own gravity makes them round
- ✔ Haven't "cleared the neighbourhood" around their orbits

I surrounded the third criterion in quotation marks because that's what the IAU said, but many astronomers don't think there's a clear explanation of what "cleared the neighbourhood" means. The general idea is that a planet's gravity disturbs objects that are in nearby orbits (but not moons of the planet) so that the objects — such as asteroids and comets — are thrown into orbits that take them away from the general vicinity of the planet. Many Kuiper Belt objects (see the final part of this chapter) are in Pluto's general vicinity, so Pluto supposedly hasn't cleared its area. But hundreds of so-called *Trojan asteroids* are right in Jupiter's orbit, and no one denies planet status to Jupiter. Did the IAU pick on Pluto because it's too small to defend itself? For more about this dispute, I immodestly recommend *Pluto Confidential: An Insider Account of the Ongoing Battles over the Status of Pluto,* by Laurence A. Marschall and me (BenBella Books, 2009). Meanwhile, keep Pluto in your thoughts, even though it's just a dwarf planet.

Figure 9-1:
Pluto is
mysterious,
rocky, and
icy.

Courtesy of NASA

Pluto is so far away that scientists have little idea about its geography. Its elongated elliptical orbit brings it within about 29.7 AU, or 2.8 billion miles (4.4 billion kilometers), of the Sun and takes it as far out as 49.5 AU, or 4.6 billion miles (7.4 billion kilometers), away.

Images from the Hubble Space Telescope show lighter and darker regions on Pluto, which may correspond to areas of fresh ice and old ice on the planet, respectively. But that's about all scientists can tell at this point. Hubble scientists mapped some of the images onto a globe of Pluto. You can see a video animation of the Pluto globe turning on its axis, to show you the whole surface of the dwarf planet. Just visit the website www.hubblesite.org/newscenter/archive/releases/2010/06/video/b/.

NASA launched the first space probe to Pluto, called New Horizons, in 2006. It will arrive at Pluto in 2015 and then head out through the Kuiper Belt, a region beyond Neptune where small, icy bodies abound (see the section "Buckling Down to the Kuiper Belt," later in this chapter, for more). You can check on the probe's progress and findings at www.nasa.gov/mission_pages/newhorizons/main/index.html.

The moon chip doesn't float far from the planet

Pluto, like Uranus, is tilted on its side. Its equator is tilted about 120° from the plane of its orbit. Astronomers assume that Pluto, like Uranus, suffered a major collision. The colliding body probably came from the Kuiper Belt. The impact tipped Pluto over, leaving it with the big tilt, and collision debris then condensed to form Pluto's large moon, Charon, much as our Moon is believed to have formed from a great impact on Earth (see Chapter 5).

Using the Hubble Space Telescope, astronomers have discovered four smaller moons of Pluto since 2004. At least three of them orbit the dwarf planet in the same plane as Charon and probably were formed from the same collision.

Pluto takes 6 days, 9 hours, and 18 minutes to turn once on its axis, and Charon orbits once around the planet in exactly the same amount of time. So the same hemispheres of Pluto and Charon always face each other. In the Earth–Moon system, one hemisphere of the Moon always faces Earth, but not vice versa. Someone standing on the near side of the Moon can see the whole planet over the course of an Earth day, but a person standing on Charon can never see more than half of Pluto.

Pluto and Charon are both icy, rocky worlds, and unlike Uranus and Neptune, their ice is real, not molten. With a surface temperature of –387°F (–233°C), almost everything freezes on Pluto. Water ice, methane ice, nitrogen ice, ammonia ice, and even frozen carbon monoxide are present on the surface. It exhausts me to think of it! Scientists have detected some, but not all, of these substances on Charon. Pluto's atmosphere contains methane, nitrogen, and carbon monoxide, and it changes with time. Since about 2000, the atmosphere has been expanding, but not uniformly in all directions. Pluto's gaseous carbon monoxide seems to be moving outward in the direction toward Charon, almost like the gas tail of a comet. (For more on comets and their tails, see Chapter 4.)

Pluto isn't uniformly arctic, however. Astronomers suspect that it has some "tropical oases" where the temperature gets all the way up to –351°F (–213°C).

Little Pluto compared to some big moons

Pluto is 1,430 miles (2,300 kilometers) in diameter, much smaller than any of the eight planets of the solar system. Pluto doesn't even measure up to the four Galilean moons of Jupiter and the moons Titan of Saturn and Triton of Neptune. (I describe each of these moons in Chapter 8.) In fact, Pluto is less than twice as big as 750-mile-wide (1,207-kilometer-wide) Charon. Before Pluto was reclassified as a dwarf planet, some astronomers called Pluto and Charon a double planet.

Buckling Down to the Kuiper Belt

Scientists estimate that about 70,000 icy bodies — called Kuiper Belt objects (KBOs) — larger than 60 miles (100 kilometers) in diameter orbit between Neptune's orbit and a distance of 50 AU from the Sun. That region is called the Kuiper Belt, after the astronomer Gerard P. Kuiper, who was one of the first scientists to predict its existence. Almost all of the KBOs are beyond the reach of backyard telescopes, unless your backyard is the site of the Palomar Observatory or a similar facility. (Amateurs with fairly large telescopes can see Pluto, which is now recognized as the first known KBO, as well as being a dwarf planet.) Astronomers David Jewitt and Jane Luu discovered the first KBO other than Pluto in 1992. Since then, researchers have found more than a thousand others.

Among the many KBOs that astronomers have discovered since 1992, a few rival Pluto in size, and at least one, Eris, may be slightly bigger. Eris, too, is a dwarf planet, and it has at least one moon, Dysnomia.

Many of the of known KBOs share three properties with Pluto:

- ✔ They have highly elliptical orbits.

- ✔ Their orbital planes are tilted by a significant angle with respect to the plane of Earth's orbit.

- ✔ They make two complete orbits around the Sun in approximately the same time that Neptune takes to make three orbits (496 years for Pluto's two orbits and 491 years for Neptune's three). This effect is called a *resonance,* and it works to keep Pluto and Neptune from ever colliding or even coming close to each other, although their orbits cross.

Pluto is safe from disturbance by the powerful gravity of the much larger Neptune, and so are the KBOs that share these three properties — called *Plutinos,* meaning "little Plutos."

Other kinds of objects are orbiting beyond Neptune and Pluto, but like the KBOs, the objects can't be very massive because their gravitational effects on known objects would give them away. One such *trans-Neptunian* body, called Sedna, was discovered in March 2004 at a distance of 90 AU from the Sun, well beyond the 50 AU distance where the Kuiper Belt seems to peter out. Sedna's size isn't accurately known, but it's probably much smaller than Pluto. Some astronomers believe that Sedna is a member of the Oort Cloud, a huge collection of distant comets that I describe in Chapter 4. The only large planets beyond Neptune are the planets of other stars (see Chapter 14).

You can find out more about KBOs on Professor David Jewitt's Kuiper Belt website at the University of California, Los Angeles (www2.ess.ucla. edu/~jewitt/kb.html).

Viewing the Outer Planets

With experience, you can locate the outer planets Uranus and Neptune, but tiny Pluto may be beyond your visual reach. The first time you look for any of these distant objects, seek the aid of a more experienced amateur astronomer, unless you have a telescope with computerized pointing. (I recommend two such telescopes, from two different manufacturers, in Chapter 3.) Even then, you will be better off with a little help from your friends.

The annual *Observer's Handbook* of the Royal Astronomical Society of Canada (www.rasc.ca) carries maps of the planets' positions during the year. Astronomy magazines (see Chapter 2) carry similar maps.

Sighting Uranus

Uranus was discovered with a telescope, and it sometimes shines bright enough to be barely visible to the eye under excellent viewing conditions. When you become an experienced observer, you'll probably be able to spot it with binoculars. Through your telescope, you can distinguish Uranus from a star, thanks to

- ✔ Its small disk, a few seconds of arc in diameter (defined in Chapter 6)
- ✔ Its slow motion across the background of faint stars

The disk of Uranus has a pale green tint; you can make out the disk with a high-power eyepiece when viewing conditions are good. (Chapter 3 has more about telescopes and eyepieces.) You can detect the motion of Uranus by making a sketch of its relative position among the stars in the field of view. For this purpose, use a low-power eyepiece so that the field of view is larger and more stars are visible. Look again in a few hours or on the following night, and sketch again.

As of March 2012, Uranus has 27 known moons. Although you can glimpse a few of the biggest moons with large amateur telescopes, they're better suited for study with powerful observatory telescopes. Uranus's dark rings are detectable with the Hubble Space Telescope and in images made with very large telescopes on Earth, but you can't see them with amateur instruments.

You can see the Hubble Space Telescope images of Uranus and its rings at www. hubblesite.org/newscenter/archive/search.php?query=uranus+ view:images. You can browse images of Uranus, its moons, and the rings from the Voyager 2 space probe at http://photojournal.jpl.nasa.gov, the Planetary Photojournal website; just click on the labeled picture of Uranus there. (Voyager 2 is the only spacecraft to have visited Uranus.)

Distinguishing Neptune from a star

Neptune appears fainter in the sky than Uranus, but it gets as bright as 8th magnitude (Chapter 1 has more about magnitudes). If Uranus challenges your observing skills, you really have to take them to the next level for Neptune when it's not at its brightest.

Neptune is about the same actual size as Uranus, but it orbits much farther away, so through a telescope, its apparent disk is smaller. You may need a large amateur telescope to discern it from a star. If you become good at perceiving pale hues in dim objects seen through a telescope, you'll be able to tell that Neptune has a blue tint.

Because Neptune orbits farther from the Sun than Uranus, it moves at a slower speed. The slower speed combined with a greater distance from Earth means that the angular rate of speed across the sky — in arc seconds per day (see Chapter 6) — is *usually* less for Neptune than for Uranus. So you may have to wait another night or two to be sure that you've seen Neptune moving across the background stars.

I say "usually" because both Uranus and Neptune, like all the planets beyond Earth's orbit, show retrograde motion at times (see Chapter 6), so they seem to slow down and reverse direction now and then. If you happen to catch Uranus when it changes direction in the sky, its apparent motion is much slower than usual; by comparison, Neptune may go full tilt at that time.

Neptune has 13 known moons, as of March 2012. The largest is Triton (see the section "Against the grain: Neptune and its biggest moon," earlier in this chapter, for more about Triton). After you master locating Neptune, look for Triton with a telescope of 6 inches or more in diameter on a clear, dark night. It has a large orbit, about 8 to 17 arc sec from Neptune (about four to eight Neptune diameters), so you may mistake Triton for a star. But by sketching Neptune and the faint "stars" around it on successive nights, you can deduce which "star" moves with Neptune across the starry background as it also moves around Neptune. It takes Triton almost six days to make one full orbit around the planet.

You can browse images of Neptune and its moons from the Voyager 2 space probe at `http://photojournal.jpl.nasa.gov`, the Planetary Photojournal website, by clicking on the Neptune link. You can see Hubble Space Telescope images at `www.hubblesite.org/newscenter/archive/search.php?query=neptune+view:images`.

Straining to see Pluto

Pluto is a much tougher viewing challenge than any planet in the solar system. It orbits far away and is small. Typically, Pluto is 14th magnitude (see Chapter 1 to find out about magnitudes). It's currently moving away from the Sun and Earth and will continue to move away for many years as it traverses its 248-year orbit.

Skilled amateurs claim to have seen Pluto with 6-inch telescopes. I recommend that you use at least an 8-inch telescope.

Pluto's largest moon, Charon, orbits very close to Pluto and revolves around it in just 6 days, 9 hours, and 18 minutes. You can distinguish it only with powerful observatory telescopes.

Part III
Meeting Old Sol and Other Stars

The 5th Wave   By Rich Tennant

"Okay—apparently this isn't the best thing to observe the sun with."

In this part . . .

*P*art III introduces you to the stars. No, not the wealthy
people in Hollywood — I'm talking about the Sun and
all the other stars in the Milky Way galaxy and beyond. You
find out about the types of stars and their life cycles, from
birth to death. When Meryl Streep, Tom Hanks, and George
Clooney are long forgotten, Alpha Centauri will still be shining. When you weary of current events, remember that the
real stars are still there for you.

I also provide a chapter on black holes and quasars, and I
simplify the subject matter so that you don't get a migraine
trying to understand it. However, the information about
space and time distortions may twist your thoughts a bit.

Chapter 10

The Sun: Star of Earth

● ●

In This Chapter

▶ Understanding the makeup of the Sun

▶ Keeping up with solar activity

▶ Viewing the Sun safely

▶ Watching sunspots and eclipses

▶ Surfing the web for solar images

● ●

Although many people are attracted to astronomy by the beauty of a moonlit night and a starry sky, you need nothing more than a sunny day to experience the full impact of an astronomical object firsthand. The Sun is the nearest star to Earth and provides the energy that makes life possible.

The Sun is so familiar in daily life that people take it for granted. You may worry about getting sunburned, but you probably seldom think of the Sun as a primary source of information about the nature of the universe. In fact, the Sun is one of the most interesting and satisfying astronomical objects to study, whether with backyard telescopes or advanced observatories and instruments in space. The Sun changes day by day, hour by hour, and moment by moment. And you can show it off to the kids without keeping them up past their bedtimes.

But don't even think about looking at the Sun, let alone showing it to a child or anyone else, without taking the proper precautions that I explain in this chapter. You don't want your view of the Sun to cost you your eyesight. Make safety in viewing your prime consideration; when you know how to protect your vision with the proper equipment and procedures, you can follow the Sun not only daily, but also over the 11-year sunspot cycle I describe later in this chapter.

This chapter introduces you to the science of the Sun, the Sun's effects on Earth and on industry, and safe solar observing. Get ready to look at the Sun in a new way — safely and with awe.

Surveying the Sunscape

The Sun is a *star,* a hot ball of gas shining under its own power with energy from *nuclear fusion,* the process by which the nuclei of simple elements combine into more complex ones. The energy produced by fusion inside the Sun powers not only the Sun itself, but also much of the activity in the system of planets and planetary debris that surrounds the Sun — the solar system of which Earth is a part (see Figure 10-1, which is not to scale).

The Sun produces energy at an enormous rate, equivalent to the explosion of 92 billion 1-megaton nuclear bombs every second. The energy comes from the consumption of fuel. If the Sun consisted of burning coal, it would burn up every last lump of itself in just 4,600 years. But fossil evidence on Earth shows that the Sun has been shining for more than 3 billion years, and astronomers are certain that it has been beaming away for longer than that. The estimated age of the Sun is 4.6 billion years, and it still burns strong today.

Only nuclear fusion can produce the Sun's huge total energy release — its *luminosity* — and keep it going for billions of years. Near the center of the Sun, the enormous pressure and the central temperature of almost 16 million degrees Celsius (29 million degrees Fahrenheit) cause hydrogen atoms to fuse into helium, a process that releases the great torrent of energy that drives the Sun.

About 700 million tons of hydrogen turn into helium every second near the center of the Sun, and 5 million tons vanish and turn into pure energy.

If humans could generate energy through fusion on Earth, all the problems with fossil fuel, including air pollution and the consumption of nonrenewable resources, would be solved. But despite decades of research, scientists still can't do what the Sun does naturally. Clearly, the Sun deserves further study.

Figure 10-1:
Pluto and
the planets
orbit the
Sun as part
of our solar
system.

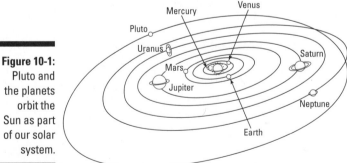

The Sun's size and shape: A great bundle of gas

When I taught Astronomy 101, I'd always pose the question, "Why is the Sun the size that it is?" Hundreds of mouths would drop open and dozens of pairs of eyes would wander the room, but hardly anyone ever had a clue. It doesn't even seem like a logical question. Everything has a size, right? So what?

But if the Sun is made of nothing but hot gas (and it is), what keeps it together? Why doesn't it all blow away, like a puffed-out smoke ring? The answer, my friend, is that gravity keeps the Sun from blowing in the wind. Gravity is the force, which I describe in Chapter 1, that affects everything in the universe. The Sun is so massive — 330,000 times the mass of planet Earth — that its powerful gravity can hold all the hot gas together.

Well, you may be wondering, if the Sun's gravity pulls all its gas together, why isn't it squeezed down into a much smaller ball? The answer is the same thing that sells many a used car: high pressure. The hotter the gas, and the more gravity (or any other force) that squeezes it together, the higher its pressure. And gas pressure inflates the Sun just like air pressure inflates an automobile tire.

Gravity pulls in; pressure pushes out. At a certain diameter, the two opposite effects are equal and in balance, maintaining a uniform size. The certain diameter is about 864,500 miles (1,392,000 km), or about 109 times Earth's diameter. You can fit 1,300,000 Earths inside the Sun, but I don't know where you'd get them.

The Sun is round for much the same reason: Gravity pulls equally in all directions toward the center, and pressure pushes out equally in all directions. If the Sun rotated rapidly, it would bulge a little at the equator and flatten slightly at the poles due to the effect that people often call *centrifugal force*. But the Sun rotates at a very slow rate — only once every 25 days at the equator (and slower near the poles) — so any midriff bulge isn't noticeable.

The Sun's regions: Caught between the core and the corona

The *photosphere* ("sphere of light") is the visible surface of the Sun (see Figure 10-2). When you glance at the bright disk of the Sun in the sky (be careful not to stare at it), you see the photosphere. When you see sunspots in a photograph of the Sun (or with a telescope as I describe later in this chapter), you're looking at a photo or a live view of the photosphere. And when I write that the diameter of the Sun is 864,500 miles, as I do in the previous section, I refer to the size of the photosphere. The temperature of the photosphere is 9,900°F (5,500°C).

Above the photosphere are the two main outer layers of the Sun:

✔ **Chromosphere ("sphere of color"):** A thin layer that you can see during a total eclipse of the Sun, when it becomes visible as a narrow red band around the dark limb of the Moon. (I describe solar eclipses in "Experiencing solar eclipses," later in this chapter.) The chromosphere measures only about 1,000 miles (1,600 km) thick, but its temperature reaches 18,000°F (10,000°C).

You can view the chromosphere at the edge of the Sun if you use an expensive H-alpha filter, mentioned in the sidebar "Solar-viewing styles of the big spenders," later in this chapter, or you can see it on images taken with professional telescopes and displayed on the NASA website (see the section "Looking at solar pictures on the Net") and on various professional observatory websites. And you may glimpse the chromosphere during a total eclipse of the Sun, covered later in this chapter. During the eclipse, you don't need equipment of any kind to see the chromosphere — just your normal vision.

The transition from the chromosphere to the hundred-times-hotter corona occurs in a very thin boundary layer called the *transition region.* It doesn't show up in views of the Sun.

✔ **Corona:** The largest and least dense (most tenuous) layer of the Sun. You see it as a pearly white region extending out from the eclipsed disk of the Sun during a total eclipse. The corona's shape is always different from day to day (as photographed by Sun-watching satellites) and from one total eclipse to the next (as seen by viewers on Earth). The corona doesn't have a particular diameter — it thins out gradually with distance from the photosphere, and the size you measure for the corona depends on the sensitivity of your measuring instrument. The more sensitive the instrument, the more of the corona you can detect. The corona is very thin and very hot — a sizzling 1.8 million degrees Fahrenheit (1 million degrees Celsius) and, in some places, even hotter.

The corona is so rarified and electrified that the Sun's magnetic field determines its shape. Where lines of magnetic force stretch and open outward into space, the coronal gas is thin and barely visible. It can readily escape in the form of solar wind (see the section "Solar wind: Playing with magnets"). Where lines of magnetic force reach up in the corona and then turn back down to the surface, they confine the coronal gas. The coronal region is thicker and brighter there. Some loop-shaped structures extend from the photosphere up into the corona and hold gas much cooler than the surroundings. These loops are called *prominences,* which you can see on the limb of the Sun during a total eclipse.

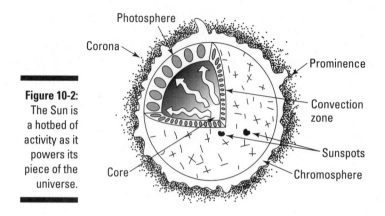

Figure 10-2:
The Sun is a hotbed of activity as it powers its piece of the universe.

Solar interior is the collective name for everything below the photosphere, and it contains the following three main regions:

- **Core:** This region extends from the center of the Sun about 25 percent of the way to the photosphere (about 108,000 miles). Solar energy is generated in the core by nuclear fusion at high temperature and high density. Because temperature and density are greatest at the center and gradually decrease with distance outward, most solar energy comes from the innermost part of the core, and less from its outer parts. The energy is generated in the form of gamma rays (a type of light) and also as neutrinos, strange subatomic particles that I describe later in this chapter in the section "Solar CSI: The mystery of the missing solar neutrinos." The gamma rays bounce off one atom to another, back and forth, but on average, they move upward and outward. The neutrinos zip right through the whole Sun and fly out into space. The farther out in the solar interior, the cooler the temperature gets.

- **Radiative zone:** This region extends from the outer boundary of the core to about 71 percent of the way from the center to the photosphere (307,000 miles from the center). This layer gets its name from the fact that much of the solar energy travels outward through the zone in the form of electromagnetic radiation (a physics term for light).

- **Convection zone:** This region begins at the top of the radiative zone, 307,000 miles from the center of the Sun, to the photosphere. Swirling currents of hot gas transfer energy in this zone, with hot currents rising from the bottom, cooling toward the top of the zone, and then falling down again, warming, and rising anew. (The same process brings heat from the bottom of a kettle of boiling water up to the surface.)

Solar activity: What's going on out there?

The term *solar activity* refers to all kinds of disturbances that take place on the Sun from moment to moment and from one day to the next. All forms of solar activity, including the 11-year sunspot cycle and some even longer cycles, seem to involve magnetism. Deep inside the Sun, a natural dynamo generates new magnetic fields all the time. The magnetic fields rise to the surface and to higher layers in the solar atmosphere. They twist around and cause sunspots, eruptions, and other changing phenomena.

Astronomers measure magnetic fields on the Sun by their effects on solar radiation, using instruments called *magnetographs*. You can see images taken with these devices on some professional solar observatory websites (see the section "Looking at solar pictures on the Net"). These magnetic field observations show that sunspots are areas of concentrated magnetic fields and that sunspot groups have north and south magnetic poles. Outside of sunspots, the overall magnetic field of the Sun is pretty weak.

Many of the rapidly changing features on the Sun and probably all explosions and eruptions seem to be related to solar magnetism. Where there are changing magnetic fields, electrical currents occur (as in a generator), and when two magnetic fields bump into each other, a short circuit — called a *magnetic reconnection* — can suddenly release huge amounts of energy.

I cover several types of solar activity in the following sections.

Coronal mass ejections: The mother of solar flares

For decades, astronomers believed that the main explosions on the Sun were *solar flares*. They thought that solar flares occurred in the chromosphere (covered earlier in this chapter) and set things off.

Now astronomers know that they were just like the blind man who feels an elephant's tail and thinks that he knows all about the beast when he's actually touching one of the animal's less significant parts. Observations from space reveal that the primary engines of solar outbursts aren't solar flares, but *coronal mass ejections* — huge eruptions that occur high in the corona. Often a coronal mass ejection triggers a solar flare beneath it in the low corona and chromosphere. You can see solar flares in many of the images on professional astronomy websites. As the number of sunspots increases over an 11-year sunspot cycle (see the following section), so does the number of flares.

Scientists didn't know about coronal mass ejections for many years because they couldn't see them. Astronomers could get a good view of the corona only at rare intervals during the brief duration of a total eclipse of the Sun (see the section "Experiencing solar eclipses," later in this chapter). But solar flares can be seen at any time, so scientists studied them intensely and overestimated their importance.

You can observe prominences (see the previous section) on the edge of the Sun even when there's no total eclipse of the Sun, but you need an expensive H-alpha filter (I describe these filters in the sidebar "Solar-viewing styles of the big spenders.") If you make enough observations, you will see that prominences sometimes erupt. These eruptive prominences may also be phases of coronal mass ejections.

When satellite images show a coronal mass ejection that isn't going off, say, to the east or to the west from the Sun, but that forms a huge expanding ring or *halo event* around the Sun, that's bad news. The halo event means that the coronal mass ejection — about a billion tons of hot, electrified, and magnetized gas — is heading right at Earth at about a million miles per hour. When it strikes Earth's magnetosphere (which I describe in Chapter 5), dramatic effects sometimes result, as I describe later in the section "Solar wind: Playing with magnets."

If you see a halo event in one of the satellite images, check the National Oceanographic and Atmospheric Administration (NOAA) Space Weather Prediction Center website (www.swpc.noaa.gov); NOAA may be forecasting some pretty fierce space weather.

Cycles within cycles: The Sun and its spots

Sunspots are regions in the photosphere where the magnetic field is strong. They appear as dark spots on the solar disk (see Figure 10-3). The spots are cooler than the surrounding atmosphere and often appear in groups.

Figure 10-3: A sunspot group 12 times bigger than Earth was photographed on September 23, 2000.

Approximate size of Earth

Courtesy of SOHO, NASA/ESA

The number of sunspots on the Sun varies dramatically over a repeating cycle that lasts about 11 years — the famous *sunspot cycle*. In the past, people blamed everything from bad weather to a decline in the stock market on sunspots. Usually, 11 years pass between successive peaks (when the most spots occur) of the sunspot cycle, but this period can vary. Furthermore, the number of spots at the peak may differ widely from one cycle to the next. Experts are always predicting how many sunspots will appear in the next cycle, but these long-range forecasts aren't very reliable.

As a sunspot group moves across the solar disk due to the Sun's rotation, the biggest spot on the forward side (the part of the group that leads the way across the disk) is called the *leading spot*. The biggest spot on the opposite end of the group is the *following spot*.

Magnetograph observations show definite patterns in most sunspot groups. During one 11-year cycle, all the leading spots in the Northern Hemisphere of the Sun have north magnetic polarity, and the following spots have south magnetic polarity. At the same time in the Southern Hemisphere, the leading spots have south polarity, and the following spots have north polarity.

Here's how these polarities are defined: The compass needle that points north on Earth is called a north-seeking compass. A north magnetic polarity on the Sun is one that a north-seeking compass would point to, if there was a compass on the Sun and it didn't melt. In the same sense, a south magnetic polarity on the Sun is one that a north-seeking compass would point away from.

Just when you think that you have it straight, guess what? A new 11-year cycle begins, and the polarities all reverse. In the Northern Hemisphere, the leading spots have south polarity, and the following spots have north polarity. In the Southern Hemisphere, the magnetic polarities reverse, too. If you were a compass, you wouldn't know whether you were coming or going.

To encompass all this information, astronomers have defined the *Sun's magnetic cycle*. The cycle is about 22 years long and contains two sunspot cycles. Every 22 years, more or less, the whole pattern of changing magnetic fields on the Sun repeats itself.

The solar "constant": Time to face the changes

The total amount of energy the Sun produces is called the *solar luminosity*. Of greater interest to astronomers is the amount of solar energy that Earth receives, or the *solar constant*. Defined as the amount of energy falling per second on 1 square centimeter of area facing the Sun at the average distance of Earth, the solar constant amounts to 1,368 watts per square meter (127 watts per square foot).

Measurements made by solar and weather satellites that NASA sent up in the 1980s revealed very small changes in the solar constant as the Sun turns. You may think that Earth receives less energy when dark sunspots are present on the solar disk, but that isn't the case. In fact, the opposite is true: more sunspots, more energy Earth receives from the Sun. Chalk up another mystery for astronomers to solve.

According to astrophysical theory, the Sun was somewhat brighter when it was very young than it has been for the last several billion years. Theory also predicts that the Sun will cast much more energy on Earth ages from now, when it becomes a red giant star (see Chapter 11).

So "solar constant" sounds like wishful thinking, although from day to day, any change in the energy output of the Sun is extremely small.

Solar wind: Playing with magnets

A type of solar plasma called the *solar wind* is always streaming out from the solar corona. It moves through the solar system at about a million miles per hour (470 kilometers per second) as it passes Earth's orbit.

The solar wind comes in streams, fits, and puffs and constantly disturbs and replenishes Earth's magnetosphere, which becomes compressed in size and swells again. The disturbances to the magnetosphere, especially those from traveling solar storms such as coronal mass ejections (covered earlier in this chapter), can cause displays of the Northern Lights (aurora borealis) and Southern Lights (aurora australis), as well as geomagnetic storms (see Chapter 5 for more on the magnetosphere and auroras). The geomagnetic storms can shut down power company utility grids (causing blackouts), blow out electronic circuits on oil and gas pipelines, interfere with GPS and radio communications, and damage expensive satellites. Some people even claim they can *hear* aurorae.

Coronal mass ejections are usually invisible with amateur equipment but marvelously revealed by satellite telescopes. They spray billion-ton blobs of electrified gas, called *solar plasma,* permeated with magnetic fields, out into the solar system, where they sometimes collide with Earth's magnetosphere. (The *magnetosphere* is a huge region around Earth in which electrons, protons, and other electrically charged particles bounce back and forth from high northern latitudes to high southern latitudes, trapped in Earth's magnetic field. The magnetosphere acts as a protective umbrella against coronal mass ejections and the solar wind.)

Solar disturbances and their effects on the magnetosphere are called *space weather.* You can see the latest official U.S. government space weather report and forecast at the website of the NOAA Space Weather Prediction Center (www.swpc.noaa.gov).

Solar CSI: The mystery of the missing solar neutrinos

The nuclear fusion at the heart of the Sun does more than change hydrogen into helium and release energy in gamma rays to heat the whole star. It also releases enormous numbers of *neutrinos,* or electrically neutral subatomic particles that have almost no mass, travel at nearly the speed of light, and can pass through almost anything. Astronomers can check on the calculated temperature and density in the core of the Sun by observing neutrinos that come from the star.

A neutrino is like a hot knife through butter: It easily cuts through. In fact, neutrinos can fly right out from the center of the Sun and into space. Those that head Earthward fly right through Earth and out the other side. But the neutrino is different from the hot knife because the knife also melts butter that it comes in contact with. The neutrino just whooshes through without affecting the matter it passes through in almost (but not quite) every case.

Physics experiments can detect the rare exceptions in which neutrinos do interact with matter, so a tiny fraction of the solar neutrinos that pass through huge underground laboratories known as neutrino observatories do get counted. These observatories are located mostly in deep mines and tunnels under mountains. At such depths, few other kinds of particles fly around, so scientists have an easier time telling a solar neutrino from other particles. One major facility, the Sudbury Neutrino Observatory in Canada, is 6,800 feet below Earth's surface. It's a good place to "delve deeply" into astronomy.

Counting neutrinos isn't easy, but some time ago, reports from the neutrino observatories indicated a deficiency in solar neutrinos: The number of neutrinos coming to Earth was significantly fewer than the number scientists expected, based on the rate at which the Sun generates energy.

The solar neutrino deficiency was the least of our problems on Earth. It paled in significance beside AIDS, war, famine, the depletion of the forests, the extinction of valuable species, and the consumption of irreplaceable fossil fuel reserves. But the loss nagged at scientists, prompting them to make new theories of particle physics and to check on theoretical models of the solar interior.

Fortunately, scientists at the Sudbury Neutrino Observatory (and elsewhere) solved the problem of the missing neutrinos. It turns out that some of the neutrinos produced in the Sun's core change to either of two other types of neutrinos on the way to Earth, and earlier neutrino observatories, which reported the solar neutrino deficiency, couldn't detect the two other types. The problem was a deficiency in laboratory equipment, not a lack of understanding of how the Sun generates energy or how many neutrinos it emits. Here's a good analogy: Suppose you have to count birds as part of an annual wildlife survey, but you wear eyeglasses with colored lenses. The colored glasses make it difficult to see birds of certain colors, so you may think that bluebirds are endangered, but the problem is that you can see only cardinals.

Four billion and counting: The life expectancy of the Sun

Someday the Sun must run out of fuel, so someday it will die. And without the energy and warmth of the Sun, life on Earth would cease to exist. The oceans would freeze, and so would the air. Seems logical, right? But what will *actually* happen is that the Sun will swell up and take the form of a red giant star (see Chapter 11 for more about red giants). It will look enormous, and it will fry the oceans. So the oceans will actually evaporate before they have a chance to freeze.

Read the preceding paragraph carefully: I didn't say that the oceans *will* freeze; I said that they would freeze without the energy of the Sun. In fact, the energy Earth receives will increase so much before the Sun dies out that humans will die of the heat (if people still exist), not of the cold. And as for the seas, boiled tuna will be served, not frozen cod. Talk about global warming!

The red giant Sun will puff off its outer layers, forming a beautiful, expanding nebula, the kind of shining gas cloud that astronomers call a planetary nebula. But no humans will be here to admire it. So to appreciate what we'll surely miss, take a good look at some of the planetary nebulas created by other stars (see Chapter 12).

The nebula will gradually fade away, and all that will remain at its center is a tiny cinder of the Sun, a hot little object called a *white dwarf star*. It won't be much larger than Earth, and although it will be very hot at first, it will be too small to cast much energy on Earth. Whatever's left on the surface of Earth will freeze. And the white dwarf will peter out like an ember in a dying campfire, gradually fading away.

Fortunately, we should have about 5 billion years to go before that prospect looms near. Future generations can worry about this, along with the national debt and how to acquire rare first editions of *Astronomy For Dummies*.

Don't Make a Blinding Mistake: Safe Techniques for Solar Viewing

Seventeenth-century Italian astronomer Galileo Galilei made the first great telescopic discovery about the Sun. By watching the daily movements of sunspots across the solar surface, he deduced that the Sun rotates. By some accounts, he damaged his eyesight while researching the Sun. Those stories may be wrong, but my warning isn't: Looking at the Sun through a telescope or another optical aid such as binoculars is extremely dangerous. A telescope or a pair of binoculars collects more light than the naked eye and focuses it on a small spot on your retina, where it will cause immediate, severe harm. Ever see a *burning glass,* a magnifying lens that focuses the rays of the Sun on a piece of paper to set it on fire? Now you get the idea.

Looking at the Sun with the naked eye isn't a good idea, and in some cases, it can be harmful. Taking even the briefest peek at the Sun through a telescope, binoculars, or any other optical instrument (whether you use your property or someone else's) is very dangerous unless the device is equipped with a properly installed solar filter made by a reputable manufacturer specifically for viewing the Sun. However, you can observe the Sun with a technique called *projection* (see the following section). If you carefully follow the instructions in the next two sections, you likely won't have a bad experience. Better yet, start your solar observing under the guidance of an experienced amateur or professional astronomer. (Head to Chapter 2 to find out about clubs and other resources that help you get going.)

Viewing the Sun by projection

Galileo invented the *projection technique* by using a simple telescope to cast an image of the Sun on a screen in the manner of a slide projector. This technique is safe only when used properly with simple telescopes, such as those sold under the description *Newtonian reflector* or *refractor.*

As I explain in Chapter 3, a Newtonian reflector uses only mirrors, aside from the eyepiece, and its eyepiece is near the top of the telescope tube, protruding at right angles. A refractor works with lenses and doesn't contain a mirror.

Don't use the projection technique with telescopes that incorporate lenses and mirrors along with eyepieces. In other words, don't use the projection technique with the Schmidt-Cassegrain or Maksutov-Cassegrain telescope models, which use both mirrors and lenses (I describe these telescopes in Chapter 3). The hot, focused solar image may damage the apparatus inside the sealed telescope tube and can pose a danger.

Here's how to safely view the Sun with the projection technique:

1. **Mount a Newtonian reflector or a refractor telescope on a tripod.**

2. **Install your lowest-power eyepiece in the telescope.**

3. **Point the telescope in the general direction of the Sun *without* sighting through or along the telescope; keep yourself and all other persons away from the eyepiece and not behind it where the focused solar beam emerges. (If the telescope has a small finder telescope mounted on it, don't look through the finder telescope either!)**

4. **Find the shadow of the telescope tube on the ground.**

Comet Hale-Bopp as it appeared in April 1997. Chapter 4 gives you the lowdown on comets, which are great blobs of ice and dust.

A view of Earth as seen from the Moon. Turn to Chapter 5 to find out more about our unique home in the solar system.

An enhanced view of the Moon, photographed by the Galileo spacecraft on its way to Jupiter. The distinct bright ray crater at the bottom of the image is the Tycho impact basin. The dark areas are impact basins filled with lava rock. Chapter 5 explores the moon in detail.

Mercury, the planet closest to the Sun, is frequently invisible to the naked eye, lost in the sun's glare. You can discover more about Mercury in Chapter 6.

Venus is covered by clouds but radar penetrates them to make this map of the planet. For more information about this dry, acidic planet, see Chapter 6.

A panorama of the Columbia Hills on Mars, from the Mars Exploration Rover Spirit. See Chapter 6 for details on Mars.

These sand
dunes are
in a crater
on Mars.

Mars is likely to
be the first planet
visited by people
from Earth.

Jupiter's four largest moons — Io, Europa, Ganymede, and Callisto — appear over the Great Red Spot in this montage. Chapter 8 probes this giant, gaseous planet and its moons.

Saturn's rings were photographed by the Hubble Space Telescope. You can see them with a small amateur telescope. Find out more about Saturn and its rings in Chapter 8.

Saturn's rings are shown here in false colors that reveal their nature. Those in turquoise are mostly ice; those in red contain many rock particles as well. The Cassini spacecraft photographed the rings in ultraviolet light, which is invisible to the human eye.

Like Saturn, Uranus has rings, but they can't be seen with a home telescope. Chapter 9 travels to the far reaches of our solar system to study Uranus.

Streaky white clouds and a large dark spot marked Neptune's atmosphere when this picture was taken. See Chapter 9 for details on Neptune.

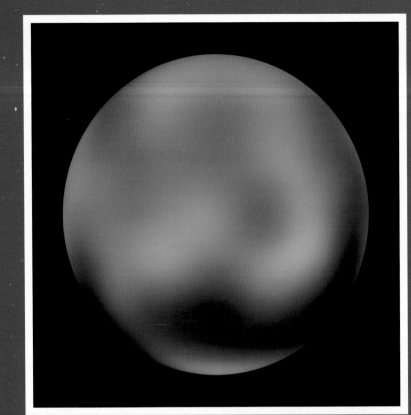

Courtesy of NASA, ESA, and M. Buie (Southwest Research Institute)

Pluto, now called a dwarf planet, has surface markings that may change with the season. To find out more about this former planet, check out Chapter 9.

Courtesy of NASA/JPL

The Galileo space probe took these nine photos as it flew past the oddly shaped asteroid Ida. Chapter 7 takes you on a tour of the Asteroid Belt.

Looking in infrared light that penetrates through dust clouds in the Milky Way, the Hubble Space Telescope photographed the Quintuplet Star Cluster near the center of our galaxy. For details on the Milky Way, see Chapter 12.

Courtesy of NASA; Don Figer, STScI

Courtesy of NASA, The Hubble Heritage Team, STScI, and AURA

The globular cluster Messler 80 is a great ball of hundreds of thousands of stars. Chapter 12 identifies the best globular clusters to view with binoculars or a small telescope.

Courtesy of NASA, Jeff Hester, and Paul Scowen, Arizona State University

The Eagle Nebula is a region in the Milky Way where new stars are being born. See Chapter 12 for details.

Courtesy of NASA, The Hubble Heritage Team, STScI, AURA

Two spiral galaxies are caught in a cosmic collision by the Hubble Space Telescope.
Chapter 12 explores the structure of spiral galaxies.

Courtesy of NASA, A. Fruchter, ERO Team, STScI

The planetary nebula
NGC 2392 in Gemini.

5. **Move the telescope up and down and back and forth while watching the shadow, to make it as small as possible.**

   The best way is for you or an assistant to hold a piece of cardboard beneath the telescope, perpendicular to the long dimension of the scope, so that the shadow of the tube falls on the cardboard. Move the telescope so that the tube shadow is as close to a solid, dark, circular shape as possible.

6. **Hold the cardboard at the eyepiece; the Sun will be in the field of view, and its image will project onto the cardboard.**

   If the Sun's image isn't in view, the bright glare of the Sun should be visible on one side of the cardboard; in that case, move the telescope to move the glare toward the center of the cardboard, thus moving the Sun into view. Keep in mind that moving the cardboard farther away from the eyepiece will make the projected image of the Sun bigger and easier to view. But if you move the cardboard too far out, the image will be too dim for viewing.

Figure 10-4 presents a diagram of the projection technique. (*Important:* The diagram shows a small finder scope mounted on the telescope because most telescopes are equipped that way, but (as I stated above) you *must not* look at the Sun through the finder scope or the main telescope because severe eye damage will occur.) The easiest and safest way to practice the technique is to consult an experienced observer from your local astronomy club; flip to Chapter 2 to find out how to locate a club in your area.

Figure 10-4:
Project the Sun onto a white surface to protect your eyes.

Dinah L. Moché/Astronomy: A Self-Teaching Guide, Seventh Edition

Even though you avoid looking through the telescope, you have to beware of some other hazards the projection method presents. I once saw a student at a school in Brooklyn project a solar image with a 7-inch telescope. He knew to avoid looking through the eyepiece. But he was careless enough to move his arm through the projected beam close to the eyepiece, where the solar image is very small. The hot concentrated image burned a smoking hole in his black leather jacket.

You must take great care when using a telescope as a solar image projector, and you must *never* allow unsupervised children or any person who isn't trained in this method to operate the telescope. Don't look at the Sun through your telescope, and don't look at the Sun through the small finder telescope or viewfinder that your telescope may be equipped with. To avoid injury, make sure that no part of anyone's body, clothing, or other property gets in the projected beam of sunlight; only your cardboard projection screen should be in the beam.

Solar-viewing styles of the big spenders

Special solar filters, called *H-alpha filters,* enable you to see some exciting solar features that you can't observe in white light. In particular, unless you are looking at a total eclipse of the Sun, these filters are necessary for viewing solar prominences, which look like fiery arches on the edge or limb of the solar disk. But these filters are very expensive (often more than $1,000).

H-alpha filters are rated according to their *bandwidth* (also called *passband*), meaning the short length of solar spectrum that they admit. Generally, the shorter the bandwidth, the more expensive the filter. Most H-alpha filters should show prominences, on the edge of the Sun, but to readily see solar flares, you need filters with bandwidths rated as 0.7 Å (0.7 angstroms) or less. You can also see *filaments* through H-alpha filters — they resemble dark, slightly curved lines on the solar disk — that are not seen with white light filters. Filaments and prominences are basically the same: They're bright in an H-alpha filter seen at the edge of the Sun against a relatively dark background,

but they're dark seen on the face of the Sun, which provides a bright background.

If the high price doesn't deter you (and you have some experience in white-light solar observing, as I describe in this chapter), be sure to investigate H-alpha filters. Thousand Oaks Optical (www.thousandoaksoptical.com) is a major source of H-alpha filters and other types of filters for solar observers. Meade Instruments sells Coronado SolarMax II filters, a high-end product that you can tune for best contrast (www.meade.com).

Perhaps the safest and least expensive way to obtain your own H-alpha-filtered views of the Sun is to use a small telescope built for that sole purpose. Specifically, the well-regarded Coronado Personal Solar Telescope, also sold by Meade, gives outstanding views. It costs about $600 without a tripod. You can see a lot of solar activity in H-alpha, but it costs much more than white-light observations, which just show sunspots.

When you get a good handle on the projection technique, you can look for sunspots. If you spot some spots, look again tomorrow and the next day to monitor their movement across the solar disk. In reality, although they may move a bit on their own, most of the sunspots' movements are due to the turning of the Sun, or the *solar rotation*. You're repeating Galileo's discovery in the safe manner that he pioneered.

If you don't want to use the projection technique, or if you have the kind of telescope that uses both lenses and mirrors — which you shouldn't use with this technique — you can still view the Sun safely, but you need a special *white light* solar filter. "White light" means that the filter admits all or most of the colors of visible light; "filter" means that it cuts down on the brightness of the light. Viewing the Sun through a safe white light filter (see the following section) requires a modest investment, but the price you pay is well worth the viewing and the safety. Read on, Galileo!

Viewing the Sun through front-end filters

The only solar filters that I recommend for observing the Sun in ordinary white light go at the *front end* of your telescope so that no light can enter the telescope without passing through the filter. (These filters are not the H-alpha filters that I describe in the sidebar "Solar-viewing styles of the big spenders," and they're much less expensive.) White light filters show sunspots, which H-alpha filters do not. But white light filters don't show solar prominences and flares, which you can see through H-alpha filters.

Filters at, near, or in place of the eyepiece may break as a result of concentrated solar heat, causing possibly great damage to your eye. Use only filters that go at the front end of your telescope.

I give you the scoop on various types of scopes in Chapter 3, and I recommend the following front-end telescope filters for solar-viewing use:

- **Full-aperture filters:** Appropriate for telescopes of 4-inch aperture or less (the *aperture* is the diameter of the light-collecting mirror or lens in your telescope), such as the Meade ETX-90 and the Celestron SkyProdigy 90 (see Chapter 3). The filter extends across the full diameter of the telescope so that the entire light-collecting mirror or lens receives the filtered light from the Sun.

- **Off-axis filters:** Best for telescopes of 4-inch aperture or more — but not for refractors. An off-axis filter is smaller than the aperture of the telescope, but it's mounted in a plate that covers the entire aperture. The Sun is so bright that you don't need the whole aperture of the telescope to collect enough light for good solar viewing. A larger aperture potentially can give a sharper view, but in most viewing locations, blurring by Earth's atmosphere nullifies this advantage. The less unneeded sunlight gets into your telescope, the safer you and the telescope will be.

Put the stop-down on your telescope

When you block some or most of the light path of a telescope (for example, by using a filter that admits light only across part of the aperture), you *stop down* the telescope.

Guess who invented the stopping down of telescopes? Galileo! What a guy! You can repeat his work by watching sunspots with a stopped-down telescope. He also conducted physics experiments, such as (allegedly) dropping weights from the Leaning Tower of Pisa. Don't even think about repeating that one.

You want an off-axis solar filter with most telescopes other than refractors because nonrefractors usually have little mirrors or mechanical devices on-center inside the telescope tube, blocking the part of the light that comes down the center of the tube.

In the special case of a refractor of 4-inch aperture or more, your filter needs to go over the top end of the telescope and be smaller than the telescope aperture, but it should mount centrally in an opaque plate that covers the telescope. The filter should mount on-center because, generally, the central part of the primary or objective lens of the refractor (the big lens) may have better optical quality than the periphery of the lens.

Thousand Oaks Optical, in Thousand Oaks, California, manufactures full-aperture and off-axis glass solar filters of several kinds. The company sells filters made to fit many different specific commercially available telescopes, as listed on its website, at http://thousandoaksoptical.com/solar.html.

Use solar filters only in accordance with the manufacturer's directions.

Fun with the Sun: Solar Observation

The Sun is a fascinating, constantly changing ball of hot gases that offers plenty of viewing opportunities for the prudent observer. In addition to observing the Sun yourself (using the precautions described in the previous section), you can visit websites that offer awe-inspiring, professionally produced pictures. And taking advantage of both provides you with the full solar-viewing experience. This section suggests some ways that you can personally enjoy old Sol.

Tracking sunspots

When you become confident in your ability to observe the Sun safely with the projection method or by equipping your telescope with a safe solar filter, you can begin to study sunspots using the following plan:

✔ Observe the Sun as often as possible.

✔ Note the sizes and positions of sunspots and groups of sunspots on the solar disk.

Some sunspots look just like tiny dark spots. If the spot is truly a tiny dark spot, even through a powerful observatory telescope, it's a *pore*. But if a sunspot is big enough, you can distinguish its different regions. The dark central portion is called the *umbra,* and the surrounding area that appears darker than the solar disk but lighter than the umbra is the *penumbra.*

✔ Chart the motion of the sunspots as the Sun makes one complete turn — which takes 25 days (at the equator) to about 35 days (near the poles; yes, the Sun turns at different rates at different latitudes, another one of its many mysterious and unexpected properties).

The Solar Section of the Association of Lunar and Planetary Observers offers a form for recording and reporting sunspots at `http://alpo-astronomy.org/solar/frm1.pdf`. The British Astronomical Association's Solar Section offers its "Grid Drawing" form for the same purpose, along with other solar observing forms. Go to `www.britastro.org`, click Solar Section, and click the link to the Solar Section website.

As you track sunspots, you may want to note how many you see in a day; that figure is called (guess what?) the *sunspot number.* You may even want to keep track of the sunspot number from year to year to see if you can measure the sunspot cycle yourself. In the following sections, I give you info on how to compute the sunspot number and where to find official numbers.

Figuring your personal sunspot number

Compute your own sunspot number for each day of observation, using this formula:

$$R = 10g + s$$

R is your personal sunspot number, g is the number of groups of sunspots you see on the Sun, and s is the total number of sunspots you count, including the spots in groups. Sunspots usually appear isolated from each other on different parts of the solar disk. Spots close together on one part of the disk are a group. A completely isolated spot counts as its own group (the reasoning behind this designation is pretty spotty, but that's the way scientists have done it for years).

Suppose that you count five sunspots; three are close together in one place on the Sun, and the other two appear at two widely separated locations. You have three groups (the group of three and the two groups consisting each of one spot), so g is 3. The number of individual spots is five, so s is 5.

$$R = (10 \times 3) + 5$$

$$R = 30 + 5$$

$$R = 35$$

Finding official sunspot numbers

On the same day, different observers come up with different personal sunspot numbers. If you have better viewing conditions and a better telescope, or maybe just a better imagination, you calculate a higher sunspot number than your neighbor Jones. You calculate $R = 59$, and that bum Jones can only claim $R = 35$. When it comes to sunspot numbers, you're way ahead! Now when it comes to whose lawn is nicer, well, I'll leave that debate up to your neighbors. "R" you clear on this?

Central authorities who tabulate and average the reports from many different observatories find by experience that some observers keep pace with Jones, some can't see as many, and some, like you, are far ahead. From this experience, the authorities calibrate each observatory or observer and make allowances in future counts so they can average the reports and get the best estimate of the sunspot number for each day.

You can check out the sunspot number every day (or whenever you want) at www.spaceweather.com.

Scientists expected the latest peak of the sunspot cycle to occur in 2012. However, it can take a few years after a peak to determine when it actually occurred. If you missed the peak, you can watch the number of spots decrease until the Sun approaches the minimum of the sunspot cycle, when you may look in vain for a decent sunspot for months at a time.

Experiencing solar eclipses

On a daily basis, the best way to see the Sun's outermost and most beautiful region, the corona, is to view the satellite images posted on the websites I list in the next section. But seeing the corona "live and in person" is a spectacle that you shouldn't deny yourself. The corona during a total eclipse of the Sun is one of nature's finest spectacles. It's why many amateur astronomers save their earnings for years to splurge on a great eclipse trip (see Chapter 2 for details). Professional astronomers make their way to the eclipse, too, even though they have satellites and space telescopes at their disposal.

The Sun experiences *partial, annular,* and *total eclipses.* The greatest spectacle is the total eclipse; some annular eclipses are well worth the trip, too. (During an annular eclipse, a thin, bright ring of the photosphere is visible around the edge of the Moon.) A partial eclipse isn't something to drive hundreds of miles out of your way to see because you don't see the chromosphere or corona, but you definitely want to check it out if one comes your way. After all, the first and last stages of a total eclipse or an annular eclipse are partial eclipses, so you need to know how to observe those stages, too.

Observing an eclipse safely

To observe a partial eclipse, or the partial eclipse phases of a total eclipse, use the solar filters I describe in the section "Viewing the Sun through front-end filters," earlier in this chapter. You can watch through binoculars or telescopes equipped with filters, you can hold a filter in front of your eyes, or you can use the technique I describe in the earlier section "Viewing the Sun by projection."

A total eclipse normally starts with a partial phase, beginning with *first contact,* when the edge of the Moon first comes across the edge of the Sun. You now see a *partial eclipse* of the Sun, signifying that you're in the *penumbra,* or light outer shadow, of the Moon (see Figure 10-5).

Figure 10-5: What happens when the Moon eclipses the Sun.

At *second contact,* the Moon's leading edge reaches the far edge of the Sun, totally blocking the Sun from your view. Now you witness a *total eclipse;* you're in the dark *umbra,* or central shadow, of the Moon. You can put down your viewing filter or your filtered binoculars and stare safely at the fantastic sight of the totally eclipsed Sun. (For complete safety, you can check out an image of a total eclipse on the cover of this book, too.) But you must not stare at the Sun when totality is over — it's unsafe to do so.

The corona forms a bright white halo around the Moon, perhaps with long streamers extending east and west. You may see thin bright polar rays in the corona off the north and south limbs of the Moon. Watch for small, bright red points, which are solar prominences that sometimes are visible to the naked eye during brief moments of the eclipse. A thin red band along the limb of the Moon is the solar chromosphere, better seen at some total eclipses than at others. Near the peak of the 11-year sunspot cycle, the corona is often round, but near the sunspot minimum part of the cycle, the corona is elongated east to west. The corona takes a different shape during every eclipse.

Not your daddy's 3-D shades

Rainbow Symphony sells Eclipse Shades that resemble the 3-D viewers you wear like eyeglasses in movie theaters. They're intended for viewing partial phases of a total eclipse or any phase of a partial solar eclipse. Check them out on the Rainbow Symphony website, at www.rainbowsymphony.com. Thousand Oaks Optical (www.thousandoaksoptical.com) offers a similar product that it calls Solar Viewers. Both types of frames are relatively inexpensive. If you take family or friends along to a solar eclipse, I suggest bringing one of these devices for each person.

Some people take the solar filters off their binoculars or telescopes and look at the totally eclipsed Sun through these instruments without the benefit of filters. This approach is very dangerous if

- You look too soon, before the Sun is totally eclipsed.
- You look too long (a very easy way to have an accident) and continue to watch through an optical instrument after the Sun begins to emerge from behind the Moon.

Be careful! As I explained in the preceding warning note, I strongly advise against telescopic and binocular viewing of the Sun without filters, even during total eclipse, unless you're viewing under the direct control of an expert. Sometimes, for example, the experienced leader of an eclipse trip or cruise group uses a public address system, computer calculations, and personal observing know-how to announce when you can look at the eclipsed Sun. The leader also tells you when you must stop, with plenty of warning.

In my experience (which was painful), the easiest way to hurt yourself is to linger looking through your binoculars or telescope for "just another second" when a tiny part of the bright visible surface of the Sun starts to emerge from behind the Moon. That tiny bright part may not make you immediately avert your vision because it doesn't seem brilliant enough. But what you don't realize is that the infrared rays (which are invisible) from the small, exposed part of the solar surface damage your eyes without dazzling them or causing immediate pain. In a few minutes or less, you begin to feel the pain. By then, the damage is done.

If you observe safely, follow all directions, and never take chances looking at the Sun, you can look forward to many happy returns of total eclipses of the sun!

Seeking shadow bands, Baily's Beads, and the diamond ring

Another good reason to avoid looking at the Sun with optical instruments during the total phase of an eclipse is that you have so much to look for all around the sky with your naked eye.

Here are some neat phenomena you can look for during an eclipse:

✔ Just before totality, so-called *shadow bands,* shimmering, low-contrast patterns of dark and light stripes, may race across the ground or, if you're at sea, across the deck of your ship. The stripes are an optical effect produced in Earth's atmosphere when the bright disk of the Sun dwindles to the last little sliver behind the eclipsing Moon before becoming completely eclipsed.

✔ *Baily's Beads* occur just instants before and after totality, when little regions of the bright solar surface shine through gaps between mountains and crater rims on the edge of the Moon. At one moment, there may be just one very bright bead. Astronomers call this aspect of the eclipse the *diamond ring.* (The bright inner corona looks like a narrow ring around the Moon, and the bright bead is the "diamond.")

✔ Wild animals (and domesticated animals, if you're near a farm) react notably to the eclipse. Birds come down to roost, cows head back for the barn, and so on. During one 19th-century eclipse, some top scientists set up their instruments in a barn, pointing the telescopes out through the door. Boy were they surprised when totality began and the livestock ran in!

✔ Where sunlight shines down through the leaves of a tree, you often see a dappled pattern on the ground below — small bright spots, shaped like the Sun. Before the eclipse begins, the dapples are round, like the uneclipsed Sun. During the partial eclipse, the dapples look like half Moons, then crescents, changing in shape according to the current phase of the eclipse. You can observe dapples like this without a tree; just bring a kitchen colander and let the Sun shine through the many little holes during an eclipse.

When the Sun becomes totally eclipsed, look at the dark sky all around the Sun. You have a rare chance to see stars in the daytime. Special articles published in astronomy magazines or posted on their websites tell you which stars and planets to look for at each total solar eclipse. Or you can figure it out yourself by simulating the sky at the date and time of the eclipse on your desktop planetarium program or similar smartphone app (see Chapter 2). All you need to do is set the program to display the sky as it will be from the place where you expect to observe.

Following the path of totality

Totality ends at *third contact,* when the Moon's trailing edge moves out across the solar disk. At the last moment of totality, a small bright area of the photosphere may emerge from behind the Moon. Then you see the *diamond ring* that I describe above. Now you're back in the penumbra, and you can see a partial eclipse. At *fourth* or *last contact,* the Moon's trailing edge moves off the forward limb of the Sun. The eclipse is over.

The whole eclipse, from first contact to last contact, may take a few hours, but the good part, totality, lasts from less than a minute to seven minutes or slightly more.

One place on the *path of totality* — the track of the center of the Moon's shadow across the surface of Earth — boasts the greatest duration. Totality is briefer everywhere else on the path. Of course, the place where the eclipse has maximum duration may not be the place where the weather prospects are best, or it may not be a place you can easily or safely reach. So advance planning of your eclipse trip is vital. At any good location, all the accommodations, rental vehicles, and so on are usually booked up at least a year or two in advance of the eclipse.

To plan your eclipse trip, pick a likely eclipse from Table 10-1 and start investigating the best way to view it.

Table 10-1	Future Total Eclipses of the Sun	
Date of Total Eclipse	*Maximum Duration (Minutes and Seconds)*	*Path of Totality*
November 13, 2012	4:02	Australia and across the South Pacific toward (but not reaching) Chile
November 3, 2013	1:40	Across the Atlantic to Africa, from Gabon to Uganda, Kenya, and Ethiopia
March 20, 2015	2:47	Across the North Atlantic south of Greenland to the Faeroe Islands, Norwegian Sea, Spitsbergen Island, and on almost to the North Pole
March 9, 2016	4:09	From the Indian Ocean across Sumatra, Borneo, the Philippine Sea, and the North Pacific
August 21, 2017	2:40	From the North Pacific across the continental United States through Oregon, Idaho, Wyoming, Nebraska, Missouri, Kentucky, Tennessee, and South Carolina, and across the Atlantic
July 2, 2019	1:02	Across the South Pacific, Chile, and Argentina
December 14, 2020	2:10	Across the South Pacific, Chile, and Argentina, and across the South Atlantic
December 4, 2021	1:54	Across the Southern Ocean, Antarctica, the Weddell Sea, the Southern Ocean again, and to the South Atlantic

A few years before each eclipse, articles with information on the weather prospects and logistics for viewing from various locations begin to appear in astronomy magazines. Check the *Sky & Telescope* and *Astronomy* magazine websites (see Chapter 2). Look for eclipse tour advertisements in the magazines and on the web. Check out the most reliable eclipse predictions on the NASA eclipse site, at `http://eclipse.gsfc.nasa.gov/eclipse.html`. See Chapter 2 for my advice on eclipse trips. And have a great time!

Looking at solar pictures on the Net

You can see current or recent professional photographs of the solar disk and sunspots (what solar astronomers call white-light photographs — white light being all the visible light of the Sun) at various places on the web. A good place to look is the site of Italy's Catania Astrophysical Observatory, `web.ct.astro.it/sun`. The white-light photo is the one labeled "Continuum," a technical term that means a colored filter wasn't used to take the picture. You can get experience in identifying a sunspot group and counting sunspots by practicing on these photos.

Sometimes the weather is cloudy in Catania, so you look elsewhere for a current professional white-light photo of the solar disk. One good place is the Full Disk Observations page of the Big Bear Solar Observatory in California, at `www.bbso.njit.edu/cgi-bin/LatestImages`. And it's never cloudy in space, so check out the current view of the Sun in white light from NASA's Solar Dynamics Observatory (SDO). Go to `http://umbra.nascom.nasa.gov/images` and click on the image labeled "continuum 4500 Å."

You can't see the corona yourself when there's no total eclipse, but you can see white-light images of the corona from SOHO, a satellite developed by the European Space Agency and NASA, at The Very Latest SOHO Images website (`http://soho.nascom.nasa.gov/data/realtime-images.html`). They're the images labeled "LASCO."

You can see recent maps of the solar magnetic field at the websites of some professional observatories, including Stanford University's Wilcox Solar Observatory (`http://wso.stanford.edu`). Another good source is the National Solar Observatory, where the magnetograms are referred to as "VSM Data" on the web page at `http://solis.nso.edu/vsm_fulldisk3.html`. (VSM is the acronym for vector spectromagnetograph, a specially designed telescope that is instrumented to measure the magnetic fields on the Sun.)

When you become an advanced astronomer and want to photograph celestial scenes through your telescope, you may want to try solar photography. You can find inspiring examples at the Mount Wilson Observatory, where researchers have been photographing the Sun since 1905. Check out the fantastic picture of an airplane silhouetted against a spotty Sun and the picture of the largest sunspot group ever photographed, from April 7, 1947. If you're lucky enough to see a sunspot group even half as large, you may be able to see it not only with a telescope, but also by looking through a solar filter without any other optical aid. The Mount Wilson site for a few historic white-light solar photographs is `http://physics.usc.edu/solar/direct.html`.

Astronomers study the Sun in all kinds of light, not just white light. Their research includes pictures taken in ultraviolet and extreme ultraviolet radiation and X-rays, which are all forms of light invisible to the eye and blocked by Earth's atmosphere. The pictures are made with telescopes mounted on satellites orbiting Earth at high altitude or taken by spacecraft located farther away and orbiting the Sun just as Earth does. Sun images from satellites and from many kinds of telescopes on the ground are available on NASA's Current Solar Images website, at `http://umbra.nascom.nasa.gov/images/latest.html`.

Another great source of solar images is the Solar Dynamics Observatory (SDO), which was launched in 2010. Check out the images and videos at the SDO website, `www.nasa.gov/mission_pages/sdo/main/index.html`. And for even more images and movies from NASA, visit the STEREO mission website, `http://stereo.gsfc.nasa.gov`.

The Sun belongs to everyone, so observe it often. You'll be glad you did!

Chapter 11

Taking a Trip to the Stars

In This Chapter

▶ Following the life cycles of the stars

▶ Sizing up stellar properties

▶ Checking out binary, multiple, and variable stars

▶ Meeting stellar personalities

▶ Stargazing and joining citizen science projects

Hundreds of billions of stars, like the Sun, make up the Milky Way galaxy (also called "the Galaxy"), where Earth resides. Likewise, the billions of other galaxies found in the universe all contain huge numbers of stars. And just like people, stars fit into dozens of classifications. The overwhelming majority fall into several simple types. These types correspond to stages in the life cycles of stars, just as you classify people by their ages.

When you understand what a star is and how it runs through its life cycle, you get a feel for these shining beacons of the night sky — and the ones that aren't so bright, too.

In this chapter, I emphasize the initial mass of a star — the mass it's born with — as the main determinant of what the star will become. I continue with the key properties of stars, along with the features of binary, multiple, and variable stars that make them so interesting for you to observe.

And no discussion of stars is complete without some gossip about the celebrities, so I introduce you to some luminaries of the night sky that you'll want to know — the leading "personalities" of the solar neighborhood.

Life Cycles of the Hot and Massive

The most important star categories correspond to successive stages in a star's life cycle: baby, adult, senior, and the dying. (What! No teenagers? The universe gave up on youth classifications after the terrible twos.) Of course, no astrophysicist worth his PhD uses such simple terms, so astronomers refer to the stages of stars as young stellar objects (YSOs), main sequence stars, red giants, and those in the end states of stellar evolution, respectively. Many stars don't really die; they continue in a new state, becoming white dwarfs, neutron stars, or black holes. But in some cases, they are completely shattered.

Here's the life cycle of a normal star with about the same mass as the Sun:

1. **Gas and dust in a cool nebula condense, forming a young stellar object (YSO).**

2. **Shrinking, the YSO dispels its remaining birth cloud, and its hydrogen "fire" ignites.**

   In other words, nuclear fusion is underway, as I explain in Chapter 10.

3. **As the hydrogen burns steadily, the star joins the main sequence.**

   I describe this stage in stellar life in the section "Main sequence stars: Enjoying a long adulthood," later in this chapter.

4. **When the star uses up all the hydrogen in its core, the hydrogen in the shell (a larger region surrounding the core) ignites.**

5. **The energy released by the burning of the hydrogen shell makes the star brighter, and it expands, which makes its surface larger, cooler, and redder. The star has become a red giant.**

6. **Stellar winds blowing off the star gradually expel its outer layers, which form a planetary nebula around the remaining hot stellar core.**

7. **The nebula expands and dissipates into space, leaving just the hot core.**

8. **The core, now a white dwarf star, cools and fades forever.**

Stars with much higher masses than the Sun have different life cycles; instead of producing planetary nebulae and dying as white dwarfs, they explode as supernovas and leave behind neutron stars or black holes. The life cycle of a massive star progresses rapidly; the Sun may last 10 billion years, but a star that begins with 20 or 30 times the Sun's mass explodes just a few million years after its birth.

Stars with much less mass than the Sun hardly have a life cycle. They begin as YSOs and join the main sequence to remain as red dwarfs forever. The explanation for this is a fundamental principle of stellar astrophysics: The bigger the mass, the fiercer and faster the nuclear fires burn; the smaller the mass, the less fiercely the fire burns, and the longer it lasts.

By the time our Sun uses up its core hydrogen, it will be at least 9 billion years old. But a red dwarf star burns hydrogen so slowly that it shines on the main sequence forever, for all practical purposes. (Given enough time, a red dwarf would exhaust its hydrogen fuel, but that amount of time is much greater than the present age of the universe, so all the red dwarfs that ever existed are still going strong now.)

The following sections describe the stellar stages in more detail.

Young stellar objects: Taking baby steps

Young stellar objects (YSOs) are newborn stars that are still surrounded or trailed by wisps of their birth clouds. The classification includes *T Tauri stars,* named for the first of their type — the star T in the constellation Taurus — and *Herbig-Haro objects,* named for the two astronomers who classified them. (Actually, H-H objects are glowing blobs of gas expelled in opposite directions from the young star, which is usually hidden from view by dust from its birth cloud.) YSOs form in stellar nurseries — called *HII regions* — such as the Orion Nebula (see Figure 11-1), where hundreds of stars have been born in the past 1 million or 2 million years.

Many of the Hubble Space Telescope images of spectacular jetlike nebulae are pictures of YSOs. The jets and other nebular surroundings are prominent, but the stars themselves are sometimes barely visible (if you can see them at all), hidden by the surrounding gas and dust. (Flip to Chapter 12 for more about nebulae.)

Figure 11-1:
The Orion Nebula cradles many young stellar objects.

Courtesy of C. R. O'Dell (Rice University) and NASA

Main sequence stars: Enjoying a long adulthood

Main sequence stars, which include our Sun, have shed their birth clouds and now shine thanks to the nuclear fusion of hydrogen into helium that goes on in the core (see Chapter 10 for more about nuclear fusion in the Sun). For historical reasons, going back to when astronomers classified stars before they understood their differences, main sequence stars are also called dwarfs (never "dwarves"). A main sequence star is a dwarf even if it has ten times more mass than the Sun.

When astronomers and science writers refer to "normal stars," they often mean main sequence stars. When they write about "Sun-like stars," they may mean main sequence stars with roughly the same mass as the Sun, give or take a factor of no more than two. The writers may also be distinguishing stars on the main sequence, no matter how massive, from stars like white dwarfs and neutron stars.

The smallest main sequence stars — much less massive than the Sun — are *red dwarfs,* which shine with a dull red glow. Red dwarfs have little mass, but they exist in enormous quantities. The vast majority of main sequence stars are red dwarfs. Like tiny gnats at the seashore, they float all around you, but you can hardly see them. Red dwarfs are so dim that you can't see even the nearest one, Proxima Centauri — which is, in fact, the nearest-known star beyond the Sun — without telescopic aid.

Red dwarfs are so much smaller, less massive, and fainter than stars like the Sun that you may be tempted to ignore them. But as mentioned earlier in this chapter, a red dwarf lasts forever, while more massive stars, like the Sun, eventually die. We may be proud of our Sun, but those puny red dwarfs will have the last laugh.

Red giants: Burning out the golden years

Red giant stars represent another kettle of starfish entirely. Red giants are much larger than the Sun. Often they measure as big around at the equator as the orbit of Venus or even the orbit of Earth. The giants represent a later stage in the life of an *intermediate-mass* star — one with several times more to somewhat less than the mass of the Sun — after it graduates from the main sequence category (see the previous section).

A red giant doesn't burn hydrogen in its core; in fact, it burns hydrogen in a spherical region just outside the core, called a *hydrogen-burning shell.* A red giant can't burn hydrogen in its core because it has already turned all of its core hydrogen into helium through nuclear fusion.

The biggest stars are the loneliest

SETI observers (the Search for Extraterrestrial Intelligence — see Chapter 14) don't point their radio telescopes at massive stars to search for radio signals from advanced civilizations. Why not? Because massive stars explode and die after lifetimes so short that scientists can't imagine intelligent life (or even primitive life) evolving on any surrounding planets before the end comes.

Massive stars are much rarer than low-mass stars. The more massive the stars, the fewer they are. So eventually, as existing stars age and the birth clouds for new stars are used up,

the Milky Way will consist overwhelmingly of just two types of stars: the red dwarfs that go on more or less forever and the white dwarfs that fade as they go. Yes, neutron stars and stellar mass black holes will dot the Milky Way, but because they represent the remains of the much rarer, more massive stars, they'll be numerically insignificant compared to the red and white dwarfs, which come from the most abundant types of main sequence stars.

Stars are like people, in that the biggest ones are rare, just as 7-foot, 5-inch basketball players are few and far between.

Stars much more massive than the Sun don't become red giants; they swell up so much that astronomers call them *red supergiants.* A typical red supergiant can be 1,000 or 2,000 times larger than the Sun and big enough to extend past the orbit of Jupiter, or even Saturn, if put in the Sun's place.

Closing time: Coming up on the tail end of stellar evolution

The *end states of stellar evolution* is a catchall term for stars whose best years are far behind them. The category encompasses

- Central stars of planetary nebulae
- White dwarfs
- Supernovas
- Neutron stars
- Black holes

These objects are all stars on their final glide paths, doomed to oblivion.

Central stars of planetary nebulae

Central stars of planetary nebulae are little stars at the centers (duh!) of a certain type of small, beautiful nebulae. (You can see a photo in the color section of this book.) These nebulae have nothing to do with planets, but in early telescopes, their images resembled greenish planets like Uranus — hence the name.

Central stars of planetary nebulae are like white dwarfs — in fact, they turn into white dwarfs. So the central stars, too, are the remains of Sun-like stars. The nebulae, each composed of gas that a star expelled over tens of thousands of years, expand, fade, and blow away. Eventually, they leave behind stars that no longer serve as the centers of anything — they become white dwarfs.

White dwarfs

White dwarfs can actually be white, yellow, or even red, depending on how hot they are. White dwarfs are the remains of Sun-like stars that take after the old generals who, according to Douglas MacArthur, never die — they just fade away.

A white dwarf is like a glowing coal from a freshly extinguished fire. It doesn't burn anymore, but it still gives off heat. White dwarfs are the most common stars after red dwarfs, but even the closest white dwarf to Earth is too dim to see without a telescope.

White dwarfs are compact stars — small and very dense. A typical white dwarf may have as much mass as the Sun, yet it takes up as much space as Earth, or a little bit more. So much matter is packed into such a small space that a teaspoon of a white dwarf would weigh about a ton on Earth. Don't try measuring it with your good silver; the spoon will get all bent out of shape.

Supernovas

Supernovas (which experts call *supernovae,* as though they all studied Latin like old-time scientists) are enormous explosions that destroy entire stars (see Figure 11-2). Various kinds of supernovas exist, but I introduce you to just the two principal varieties.

The first type you need to know about is Type II (hey, I didn't invent the numbering system). A *Type II supernova* is the brilliant, catastrophic explosion of a star much larger, brighter, and more massive than the Sun. Before the star exploded, it was a red supergiant or maybe was even hot enough to be a blue supergiant. Regardless of color, when a supergiant explodes, it may leave behind a little souvenir, which is a neutron star. Or much of the star may implode (fall in on its own center) so effectively that it leaves behind an even weirder object, a black hole.

The second important type of supernova is called a Type Ia. *Type Ia supernovas* are even brighter than Type IIs, and they explode in a reliable manner. The actual brightness, or luminosity, of a Type Ia is always about the same; therefore, when astronomers observe a Type Ia supernova, we can figure out how far away it is by how bright it appears to us on Earth. The farther away it is, the dimmer the supernova looks. Astronomers use Type Ia supernovas to measure the universe and its expansion. In 1998, two groups of astronomers studying Type Ia supernovas discovered that the expansion of the universe isn't slowing down — it's expanding at an ever-faster rate. This discovery was contrary to previous belief, so it led experts to revise their theories of cosmology and the Big Bang and to recognize the existence of the mysterious *dark energy* (I describe dark energy and the Big Bang in Chapter 16).

Type Ia supernovas all produce similar explosions because they erupt in binary systems (covered later in this chapter) in which gas from one star flows down onto the other (a white dwarf), building up an outer hot layer that reaches a critical mass and then explodes, shattering the white dwarf. With less than critical mass, no explosion occurs. With critical mass, a standard explosion results. With more than the critical mass . . . wait — you can't have more than critical mass because the star already exploded! Astrophysics isn't so hard, is it?

Experts have argued for years over the kind of binary system that produces a Type Ia supernova. According to one theory, the two stars in the binary are a white dwarf and a larger star, like the Sun. The white dwarf sucks gas off its bigger partner. Another leading theory proposes that both stars in the binary system are white dwarfs. As of 2012, it looks like both theories are correct: Some supernovae come from the big star/little star kind of system, and others come from a pair of equals.

Figure 11-2:
A supernova in the spiral galaxy M51.

Supernova Near Nucleus of Galaxy M51

Hubble Space Telescope · Wide Field Planetary Camera 2

Courtesy of NASA

Neutron stars

Neutron stars are so small that they look up to white dwarfs, but they also outweigh them. (More accurately, they "outmass" them. Weight is just the force that a planet or other body exerts on an object of a given mass. You would weigh different amounts on the Moon, Mars, and Jupiter than you do on Earth, even though your mass stays the same — unless you overeat or go on a crash diet.)

Neutron stars are like Napoleon: small in stature, but not to be underestimated. (Figure 11-3 features a neutron star.) A typical neutron star spans only one or two dozen miles across, but it has half again or even twice the mass of the Sun. A teaspoon of neutron star material would weigh about a billion tons on Earth.

Some neutron stars are better known as *pulsars.* A pulsar is a highly magnetized, rapidly spinning neutron star that produces one or more beams of radiation (which may consist of radio waves, X-rays, gamma rays, and/or visible light). As a beam sweeps past Earth like a searchlight from a galactic supermarket opening, our telescopes receive brief spurts of radiation, which we call "pulses." So guess how pulsars get their name. Your pulse rate tells you how rapidly your heart is beating. A pulsar's rate tells how fast it spins. The rate can be a few hundred times per second or just once every few seconds.

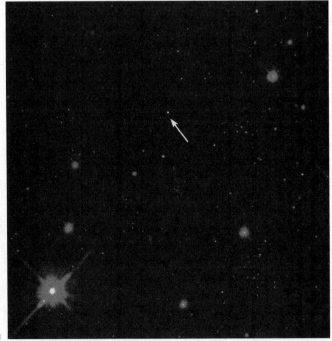

Figure 11-3:
A neutron star (at the arrow) photographed by the Hubble Space Telescope.

Courtesy of Fred Walter (State University of New York at Stony Brook) and NASA

Black holes

Black holes are objects so dense and compact that they make neutron stars and white dwarfs seem like cotton candy. So much matter is packed into such a small space in a black hole that its gravity is strong enough to prevent anything, even a ray of light, from escaping. Physicists theorize that the contents of a black hole in effect have left our universe. If you fall into a black hole, you can kiss your universe good-bye.

You can't see the light from a black hole because the light can't get out, but scientists can detect black holes by their effects on surrounding objects. Matter in the vicinity of a black hole gets hot and rushes madly around, but it never gets organized; instead, the powerful gravity of the black hole pulls matter into it, and "that's all folks."

Actually, I oversimplified; a bit of the matter swirling around the black hole does escape — just in time sometimes. The hole shoots it out in powerful jets at a significant fraction of the speed of light (which is 186,000 miles per second in a vacuum such as outer space).

This is how scientists detect black holes:

- ✔ Gas swirls around that's just too hot for normal conditions.

- ✔ Jets of high-energy particles make their escape and avoid falling into the black hole.

- ✔ Stars race around orbits at fantastic speeds, driven by the gravitational pull of an enormous unseen mass.

Astronomers have found lots of evidence for two basic kinds of black holes and limited, but increasing, information on a third kind. Here's a rundown of the three types:

- ✔ **Stellar mass black holes:** These black holes have — you guessed it — the mass of a star. More accurately, these black holes range from about three times the mass of the Sun up to perhaps a hundred times the solar mass, although astronomers haven't found any as heavy as that. Stellar mass black holes are about the size of neutron stars. A black hole with ten times the mass of the Sun has a diameter of about 37 miles (60 km). If you could squeeze the Sun down to a size compact enough to make it a black hole (fortunately, this is probably impossible), its diameter would be 3.7 miles (6 km). Stellar mass black holes form in supernova explosions and possibly by other means.

- ✔ **Supermassive black holes:** These monsters have masses of hundreds of thousands to more than 20 billion times the mass of the Sun (for some examples, see Table 13-1, "Black Hole Measurements"). Generally, supermassive black holes are located at the centers of galaxies. I want to say that they "gravitate" there, but most likely they form right there, or the galaxy forms around them. The Milky Way Galaxy has a central black hole, known as Sagittarius A*. (Nope, the asterisk doesn't refer you to a footnote. When saying the name out loud, you say *Sagittarius A star.*) It weighs in at about 4 million solar masses, and we in the solar system orbit around that black hole once every 226 million years — the latest value from the Very Long Baseline Array, a radio telescope with component antennas that stretch across U.S. territory from the Virgin Islands, through North America, and out to Hawaii. Astronomers think that a supermassive black hole exists at the center of every galaxy, or at least the center of every full-size galaxy. We're not so sure about dwarf galaxies. (Get the scoop on galaxies in Chapter 12.)

When I talk about the size of a black hole, I mean the diameter of its *event horizon.* The event horizon is the spherical surface around the black hole where the velocity needed for something to escape the black hole equals the velocity of light. Outside the event horizon, the escape velocity is smaller, so light or even high-speed matter can escape.

✔ **Intermediate mass black holes (IMBHs):** IMBHs are a poorly understood class of black holes, meaning that astronomers don't know what they are. They have estimated masses that range from a few hundred to ten thousand or more times the mass of the Sun. An IMBH is more massive than any known star, so it probably wasn't formed by the collapse of a single star (which is how stellar mass black holes are formed). On the other hand, IMBHs are found outside the central regions of galaxies, while supermassive black holes are always found in galaxy centers, where they surely formed. So IMBHs are not made where supermassive black holes form, and they are not made in the collapse of a star, as is a stellar mass black hole. What creates an IMBH? Inquiring minds want to know, but I don't recommend looking for the answer in the *National Enquirer.*

To tell the truth, supermassive black holes aren't stars. Most likely, neither are intermediate mass black holes. But I have to mention them someplace! You can't call yourself an astronomer if you don't know about black holes. (Check out Chapter 13 for even more about them.) If you pass yourself off as an astronomer, folks will ask you all kinds of questions about black holes. But how many questions do you think you'll get about main sequence stars and young stellar objects?

Star Color, Brightness, and Mass

The significance of the different types of stars (see the section "Life Cycles of the Hot and Massive") becomes clearer when you see basic observational data plotted on an astrophysicist's graph. The data are the magnitudes (or brightnesses) of the stars, which are plotted on the vertical axis, and the colors (or temperature), which are marked on the horizontal axis. This graph is called a *color-magnitude diagram,* or the Hertzsprung-Russell or H-R diagram, after the two astronomers who first made such diagrams (see Figure 11-4).

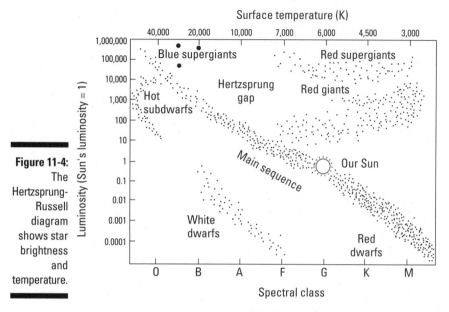

Surface temperature (K)

Figure 11-4:
The
Hertzsprung-
Russell
diagram
shows star
brightness
and
temperature.

Dinah L. Moché/Astronomy: A Self-Teaching Guide, Seventh Edition

As a college astronomy teacher, I could always tell who studied and who didn't. When I asked what data are plotted on the H-R diagram, the students who answered "H and R" revealed that they were just guessing.

Spectral types: What color is my star?

Hertzsprung and Russell didn't have good information on the colors or temperatures of stars, so they plotted the spectral type on the horizontal axis of their original diagrams. The *spectral type* is a parameter assigned to a star on the basis of its spectrum. The *spectrum* is the way the light of a star appears when a prism or other optical device in an instrument called a spectrograph spreads it out.

At first, astronomers didn't have a clue about what spectral types meant, so they just grouped stars together (under the headings of Type A, Type B, and so on) based on similarities in their spectra. Later some astronomers realized that the spectral types reflect the temperatures and other physical conditions in the atmospheres of the stars, where their light emerges into space. When scientists understood what the colors signified, they organized the spectral types in temperature order, and Hertzsprung and Russell plotted them on their diagrams. Some of the original types were deemed superfluous and dropped.

The main spectral types on the H-R diagram are O, B, A, F, G, K, and M, going from the hottest stars to the coolest. College students memorize this sequence with the help of mnemonic devices such as "Oh, be a fine girl (guy), kiss me." Table 11-1 describes the general properties of stars of each spectral type.

Table 11-1		Spectral Types of Stars	
Type	**Color**	**Surface Temperature**	**Example**
O	Violet-white	30,000°K or more	Lambda Orionis
B	Blue-white	12,000°K to 30,000°K	Rigel
A	White	8,000°K to 12,000°K	Sirius
F	Yellow-white	6,000°K to 8,000°K	Procyon
G	Yellower-white	5,000°K to 6,000°K	Beta Aquilae
K	Orange	3,000°K to 5,000°K	Arcturus
M	Red	Less than 3,000°K	Antares

Star light, star bright: Luminosity classifications

The spectral types O, B, A, F, G, K, and M have subdivisions denoted by Arabic numerals, which allow astronomers to classify stars in finer detail. Each spectral letter type has up to ten subdivisions. For example G stars encompass ten types, from G0 to G9. The lower the subdivision number, the hotter the star. The Sun's spectral type is G2, while Beta Aquilae (listed in Table 11-1) has spectral type G8. In other words, the Sun is hotter than Beta Aquilae, and the latter is almost cool enough to fall in spectral type K.

I tell you all this so that when you look up a star like the Sun or Beta Aquilae in an astronomy book (or other resource, such as the Internet), you know what it means when it lists the Sun as G2 or Beta Aquilae as G8.

But wait, there's more: Some books just use G2 or G8 for these two stars, but others describe the Sun and other stars with an additional tag, usually a Roman numeral. So you may see the Sun called a G2 V star and Beta Aquilae referred to as G8 IV. Astronomers call the Roman numeral the star's *luminosity class.*

Spectral type, like G2, refers to the temperature of a star, while luminosity class, like IV or V, describes its size and also its average density (because larger stars usually have lower density than small stars). I summarize stars by luminosity class and size in Table 11-2.

Table 11-2	Luminosity Classes for Stars	
Class	*Description*	*Example*
I	Supergiant	Rigel
II	Bright giant	Gamma Aquilae
III	Giant	Aldebaran
IV	Subgiant	Beta Aquilae
V	Main sequence (dwarf)	Rigel Kentaurus
D	White dwarf	Sirius B

You will occasionally find a star listed as luminosity class Ia or Ib. These designations give astronomers more detail: *Ia* refers to a brighter supergiant, and *Ib* indicates a fainter supergiant. But any supergiant is much brighter than the stars of the other luminosity classes.

D, the luminosity class for white dwarfs, could have been defined as the Roman numeral *D,* meaning 500. But it's actually just the English letter *D,* short for "dwarf." After you've mastered the subject of spectral types and luminosity classes, you can flash a *V* for victory, but to an astrophysicist, it's a also a *V* for main sequence dwarf.

The brighter they burn, the bigger they swell: Mass determines class

A star with greater mass concentrates a fiercer nuclear fire in its core and produces more energy than a star of lower mass, so a more massive main sequence star is brighter and hotter than a less massive main sequence star. The more massive stars usually are bigger, too. With this information, you can follow the fundamental point of stellar astrophysics reflected in the H-R diagram: Mass determines class.

On the H-R diagram (see Figure 11-4), magnitude (or luminosity) is plotted with greater luminosities or brighter magnitudes higher on the graph, and spectral type is plotted with the hotter stars to the left and the cooler stars to the right. Temperature runs from right to left, and magnitude runs from top to bottom.

On any H-R diagram, the plot of real observational data, where each point represents a single star, reveals a great deal to the careful reader:

✔ Most stars are on a band that runs diagonally from upper left to lower right. The diagonal band represents the main sequence, and all the stars on the band are normal stars like the Sun, burning hydrogen in their cores.

✔ Some stars are on a wider, sparser, and roughly vertical band that stretches up and a little to the right from the diagonal band toward brighter magnitudes (higher luminosities) and cooler temperatures. This band is the *giant sequence* and consists of red giant stars.

✔ A few stars are located all across the top of the diagram, from left to right. These stars are supergiants; blue supergiants are on the left side of the diagram, more or less, and red supergiants (which outnumber the blue) are on the right side.

✔ A few stars are located far below the diagonal band, down at the bottom left and bottom center of the diagram. These stars are white dwarfs.

Astronomers plot a main sequence star on the H-R diagram according to its brightness and temperature, but its brightness and temperature depend on only its mass. The diagonal shape of the main sequence represents a trend from high-mass to low-mass stars. The stars on the upper left of the main sequence have higher masses than the Sun, and the stars on the far right have lower masses.

Astronomers usually don't plot young stellar objects on the same H-R diagram with all the other stars, but if they did, they would plot the YSOs on the right side of the diagram, above the main sequence — but not nearly as high up as the supergiant stars. Neutron stars and black holes are too dim to plot on the same H-R diagrams with normal stars.

The H-R diagram

With just a little more explanation, you, too, can be a stellar astrophysicist and understand in one fell swoop why stars fall into different parts of the H-R diagram. Researchers spent decades figuring this out, but I give it to you on a plate. To keep it simple, I discuss a calibrated H-R diagram, where all the stars are plotted according to true brightness.

Consider this: Why is one star brighter or dimmer than another? Two simple factors determine the brightness of a star: temperature and surface area. The bigger the star, the more surface area it has, and every square inch (or square centimeter) of the surface produces light. The more square inches, the more light. But what about the amount of light a given square inch produces? Hot objects burn brighter than cool objects, so the hotter the star, the more light it generates per square inch of surface area.

Got all that? Here's how it all goes together:

✔ **White dwarfs** are near the bottom of the diagram because of their small size. With very few square inches of surface area (compared to stars like the Sun), white dwarfs just don't shine as brightly. As they fade away like old generals, they move down the diagram (because they get dimmer) and farther to the right (because they get cooler). You don't see many white dwarfs on the right side of the H-R diagram because the cool white dwarfs are so faint that they usually fall below the bottom of the diagram as printed in books.

✔ **Supergiants** are near the top of the H-R diagram because they're super big. A red supergiant can be more than 1,000 times larger than the Sun (if you put a supergiant in the Sun's place, it would extend beyond Jupiter's orbit). With all that surface area, supergiants are naturally very bright.

The fact that the supergiants are roughly at the same height on the diagram from left to right indicates that the blue supergiants (the ones on the left) are smaller than the red supergiants (the ones on the right). How do you know that? Supergiants are blue because they burn hotter, and if they burn hotter, they produce more light per square inch. Because their magnitudes are roughly the same (all the supergiants are near the top of the diagram), the red supergiants must have larger surface areas to produce equal total light (because they produce less light from each square inch).

✔ **Main sequence** stars are on the diagonal band that runs from the upper left to the lower right on the diagram, because the main sequence consists of all stars that burn hydrogen in their cores, regardless of their size. But a difference in size affects where the main sequence stars appear on the H-R diagram. The hotter main sequence stars (the ones on the left) are also bigger than the cool main sequence stars, so hot main sequence stars have two things going for them: They have larger surface areas, and they produce more light per square inch than the cool stars. The main sequence stars at the far right are the dim and cool red dwarfs.

The brown dwarfs don't top the charts

Brown dwarfs — discovered in the mid-1990s — are recent additions to the celestial inventory. They're smaller and less massive than stars and about as big as a gas giant planet like Jupiter, but they're much more massive than Jupiter. (Brown dwarf masses range from about 13 to about 70 times the mass of Jupiter.) They shine by their own light, like a star, not by reflected light, like Jupiter. But brown dwarfs aren't true stars because nuclear fusion operates only briefly in their cores. After fusion stops, they generate no more energy and just cool and fade. Their spectral types range from the cool end of type M to the successively cooler types, L and T. (Astronomers suspect that there may be even cooler brown dwarfs; if so, they would belong to spectral class Y.) On the H-R diagram in Figure 11-4, brown dwarfs fall at the extreme bottom right or even just off the chart to the right of the bottom-right corner.

Eternal Partners: Binary and Multiple Stars

Two stars or three or more stars that orbit around a common center of mass are called binary stars or multiple stars, respectively. Studies of binary and multiple stars help scientists understand how stars evolve. These small stellar systems are also fun to observe with backyard telescopes.

Binary stars and the Doppler effect

About half of all stars come in pairs. These *binary stars* almost always are *coeval,* a fancy term for "born together." Stars that form together, united by their mutual gravity as they condense from their birth clouds, usually stay together. What gravity unites, other celestial forces rarely can break apart. A grown star in a binary system has never had another partner (well, hardly ever — some strange cases occur in dense star clusters, where stars can come so close that they may actually lose or gain a partner).

A binary system consists of two stars that each orbit a common *center of mass.* The center of mass of two stars that have exactly equal masses falls exactly halfway between them. But if one star has twice the mass of the other star, the center of mass is closer to the heavier star; in fact, the center is twice as far from the lighter star as from the heavier star. If one star has one-third the mass of its heavy companion, it orbits three times farther from the center of mass, and so on. The two stars are like kids on a seesaw: The heavier kid has to sit closer to the pivot so the two are in balance.

The two stars in a binary system follow orbits equal in size if the stars are equal in mass. Stars with different masses follow orbits of different sizes. The general rule: The big guys follow smaller orbits. You may think that binary systems are like our solar system, where the closer a planet orbits to the Sun, the faster it goes and the less time it takes to make one complete orbit. That idea may be a reasonable one, but it's wrong nonetheless.

In binary systems, the big star that follows the smaller orbit travels slower than the little star in a big orbit. In fact, their respective speeds depend on their respective masses. The star that carries one-third the mass of its companion moves three times as fast. By measuring orbital velocity, astronomers can determine the relative masses of a binary system's members.

The fact that the orbital speeds of the member stars of a binary system depend on mass is what makes binary stars attract the high interest of astronomers. If one star is three times more massive than the other, it moves around its orbit in the binary star system at one-third the orbital speed of its companion star. All astronomers have to do to figure out the stars' relative

masses (meaning how much more massive one star is than the other) is measure their velocities. However, only rarely can astronomers track the stars as
they move because most binary stars are so far away that we can't see them
moving around their orbits. Fortunately, instead of throwing up their hands
in defeat, astronomers have been able to measure star masses by studying
the light from a binary star and analyzing its spectrum — a spectrum that
may be the combined light of both stars in the binary.

A phenomenon called the *Doppler effect* helps astronomers figure out the
masses of binary stars by studying their stellar spectra.

Here's all you need to know about the Doppler effect, named for Christian
Doppler, a 19th-century Austrian physicist:

The frequency or wavelength of sound or light, as detected by the observer,
changes, depending on the speed of the emitting source with respect to the
observer. For sound, the emitting source may be a train whistle. For light,
the source may be a star. (Higher-frequency sounds have a higher pitch; a
soprano has a higher pitch than a tenor.) Higher-frequency light waves have
shorter wavelengths, and lower-frequency light waves have longer wavelengths. In the simple case of visible light, the shorter wavelengths are blue
and the longest wavelengths are red.

According to the Doppler effect:

✔ If the source is moving toward you, the frequency that you detect or
  measure gets higher, so

  • The pitch of the train whistle seems to be higher.

  • The light from the star seems to be bluer.

✔ If the source is moving away, the frequency gets lower, so

  • The whistle you hear has a lower pitch.

  • The star looks redder.

The train whistle is the official example of the Doppler effect that instructors
have taught to generations of sometimes unwilling high school and college
students. But who listens to train whistles anymore? A more familiar analogy
is the way you feel the waves on the water as you zip around in a motorboat.
As you ride toward the direction from which the waves are coming, you feel
the boat rocked rapidly by choppy waves. But when you head back toward
the beach, the rocking becomes much gentler and the same waves are less
choppy. In the first case, you moved toward the waves, meeting them before
they would've met you if you stood (or floated) still. So the frequency at
which the waves struck the boat was greater than if the boat had remained
at rest. The frequency of the waves doesn't change, but the frequency of the
waves that you sense changes.

The spectrum of a star contains some dark lines — places (wavelengths or colors of light) where the star doesn't emit as much light as at adjacent wavelengths. The decreased emission at those wavelengths is caused by the absorption of light by particular kinds of atoms in the atmosphere of the star. The dark lines form recognizable patterns, and when the star is moving back and forth in its orbit, the Doppler effect makes the patterns of lines move back and forth in the spectrum detected on Earth. When the spectral lines shift toward longer wavelengths, the phenomenon is a *redshift.* When they shift toward shorter wavelengths, it's a *blueshift.* Other ways of producing redshifts and blueshifts exist, but the Doppler effect is the most familiar cause.

So by observing the spectra of binary stars and seeing how their spectral lines shift from red to blue to red again as the stars orbit, astronomers can determine their velocities and, therefore, their relative mass. And by seeing how long it takes for a spectral line to go as far to the red as it goes and then how long it takes to go as far to the blue as it goes and back to the red again, astronomers can tell the duration or period of the binary star orbit.

If you know that the period of one complete orbit is 60 days, for example, and you know how fast the star is moving, you can figure out the circumference of the orbit and, thus, the radius of its orbit. After all, if you drive nonstop from New York City to a town in upstate New York in three hours (good luck with the traffic!) at 60 miles per hour, you know that the distance you travel is 3×60, or 180 miles.

Two stars are binary, but three's a crowd: Multiple stars

Double stars are two stars that appear close to each other as seen from Earth. Some double stars are true binaries, orbiting their common centers of mass. But others are just *optical doubles,* or two stars that happen to be in nearly the same direction from Earth but at very different distances. They have no relation to each other; they haven't even been introduced.

Triple stars are three stars that appear to be close but, like the members of a double star, may or may not be all that close. But a *triple-star system,* like a binary system, consists of three stars held together by their mutual gravitation that all orbit a common center of gravity.

A comparison to wedded (or unwedded) bliss may be in order. "Three's a crowd" is an expression of the instability in most romantic arrangements when a third person becomes involved. The same is true of triple-star systems: They consist of a close pair or binary system and a third star in a much bigger orbit. If all three stars moved on close orbits, they would interact gravitationally in chaotic ways, and the group would break up, with at least one star flying away, never to return. So a triple system is effectively a "binary star" in which one member is actually a very close star-pair.

TECHNICAL STUFF

Stellar spectroscopy in a nutshell

Stellar spectroscopy is the analysis of the lines in the spectra of stars and is by far an astronomer's most important tool for investigating the physical nature of stars. Spectroscopy reveals

✔ Radial velocities (motions toward or away from Earth) of stars

✔ Relative masses, orbital periods, and orbit sizes of stars in binary systems

✔ Temperatures, atmospheric densities, and surface gravities of stars

✔ Magnetic fields and their strengths on stars

✔ Chemical composition of stars (what atoms are present and in what states they exist)

✔ Sunspot cycles of stars (well, starspot cycles)

All this information comes from measuring the positions, widths, and strengths (how dark or how bright they are) of the little dark (or sometimes bright) lines in the spectra of stars. Scientists analyze them with the help of the Doppler effect to find out how fast stars move, the size of their orbits, and their relative masses. Other phenomena, such as the *Zeeman effect* and the *Stark effect,* affect the appearance of the spectral lines. Applying their knowledge of these effects, astronomers can figure the strength of the magnetic field of the star from the Zeeman effect and determine the density and surface gravity in the star's atmosphere from the Stark effect. The very presence of particular spectral lines, each of which comes from a specific kind of atom that's absorbing (dark lines) or emitting (bright lines) light in the atmosphere of a star, tells astronomers about some of the chemical elements present in the

atmosphere of the star and the temperatures in the star where those atoms are emitting or absorbing the light.

The spectral lines even tell astronomers what condition or *ionization state* the atoms are in. Stars are so hot that the heat may strip atoms of iron, for example, of one or more of their electrons, making them iron ions. Each type of iron ion, depending on how many electrons it has lost, produces spectral lines with different characteristic patterns and positions in the spectrum. By comparing the spectra of stars recorded with telescopes with the spectra of chemical elements and ions as measured in laboratory experiments or calculated with computers, astronomers can analyze a star without ever coming within light-years of it.

In cool stellar gases, much of the iron loses only one electron per atom, so it produces the spectrum of singly ionized iron. But in the very hot parts of stars, such as the million-degree corona of the Sun, iron may lose ten electrons; the element is in a high-ionization state, and it produces the corresponding pattern of spectral lines. That pattern clearly points to the existence of a very high temperature region on the star.

Certain parts of the Sun's spectrum change along with the coming and going of disturbed regions on the Sun, which peak about every 11 years (as I explain in Chapter 10). Similar changes occur in the spectra of other Sun-like stars. So astronomers can tell the length of the sunspot cycle of a distant star by using spectroscopy, even though the star is too far away to catch a glimpse of its sunspots.

Quadruple stars are often "double-doubles," consisting of two close binary star systems that each revolve around the common center of mass of the four stars.

Multiple stars is the collective term for star systems larger than binaries: triples, quadruples, and more. At some point, the distinction between a large multiple-star system and a small star cluster becomes blurred. One is essentially the same as the other. (I describe star clusters in Chapter 12.)

Change Is Good: Variable Stars

Not every star is, as Shakespeare wrote, as "constant as the Northern Star." In fact, the North Star isn't constant, either. The famous star, also called Polaris, is a *variable star,* meaning one whose brightness changes from time to time. For many years, astronomers thought that they had the North Star's brightness changes down pat. It seemed to brighten a little and fade a little over and over, reproducibly. But suddenly its expected changes, well, changed. This difference in the pattern may signify a physical change over time, and scientists are studying what it means. Recently, astronomers at Villanova University concluded that the North Star has brightened by about 1 magnitude (about 2.5 times) since antiquity.

Variable stars come in two basic types:

✔ **Intrinsic variable stars:** These stars change in brightness because of physical changes in the stars themselves. These stars divide into three main categories:

- Pulsating stars

- Flare stars

- Exploding stars

✔ **Extrinsic variable stars:** These stars seem to change in brightness because something outside the star alters its light, as visible from Earth. The two main types of extrinsic variable stars are

- Eclipsing binaries

- Microlensing event stars

I describe each of these basic types of variable stars in the following section.

Go the distance: Pulsating stars

Pulsating stars bulge in and out, getting bigger and smaller, hotter and cooler, brighter and dimmer. These stars simply oscillate like throbbing hearts in the sky.

Cepheid variable stars

The most important pulsating stars, from a scientific standpoint, are the Cepheid variable stars, named for the first studied star of their type, Delta in the constellation Cepheus (Delta Cephei).

American astronomer Henrietta Leavitt discovered that Cepheids have a *period-luminosity relation.* This term means that the longer the period of variation (the interval between successive peaks in brightness), the greater the true average brightness of the star. So an astronomer who measures the apparent magnitude of a Cepheid variable star as it changes over days and weeks, and who thereby determines the period of variability, can readily deduce the true brightness of the star.

Why do astronomers care? Well, knowing the true brightness enables us to determine the distance of the star. After all, the farther the star, the dimmer it looks, but it still has the same true brightness.

Distance dims stars according to the *inverse square law:* When a star is twice as far away, it looks 4 times as faint; when the distance is tripled, the star looks 9 times as faint; and when a star is 10 times farther away, it looks 100 times as faint.

The headlines about the Hubble Space Telescope determining the distance scale and age of the universe came from a Hubble study of Cepheid variable stars. Those Cepheids are in faraway galaxies. By tracking their brightness changes and by using the period-luminosity relation, the Hubble observers figured out just how far away the galaxies are.

RR Lyrae stars

RR Lyrae stars are similar to Cepheids, but not as big and bright. Some RR Lyrae stars are located in globular star clusters in our Milky Way, and they have a period-luminosity relation, too.

Globular clusters are huge balls of old stars that were born while the Milky Way was still forming. A few hundred thousand to a million or so stars are all packed in a region of space only 60 to 100 light-years across. Observing the changes in brightness of RR Lyrae stars enables astronomers to estimate their distances, and when the stars are in globular clusters, it tells us how far away the clusters are. (For more on globular and other star clusters, turn to Chapter 12.)

Why is it so important to know the distance of a star cluster? Here goes: All the stars in a single cluster were born from a common cloud at the same time, and they're all at nearly the same distance from Earth because they exist in the same cluster. So when scientists plot the H-R diagram of stars in a cluster, the diagram is free of errors that may be caused by differences in the distances of the stars. And if scientists know the distance of the cluster, they can convert all the plotted magnitudes to actual luminosities, or the rates at which the stars produce energy per second. Such quantities can be directly compared with astrophysical theories of the stars and how they generate their energy. This stuff keeps astrophysicists busy.

Long period variable stars

Astrophysicists celebrate the information gleaned from Cepheid and RR Lyrae variable stars. Amateur astronomers, on the other hand, delight in observing long period variables, also called Mira stars (or Mira variables). Mira is another name for the star Omicron Ceti, in the constellation Cetus (the Whale), the first-known long period variable star.

Mira variables are huge red stars that pulsate like Cepheids, but they have much longer periods, averaging ten months or more, and the amount by which their visible light changes is even greater. At its brightest, Mira itself is visible with the naked eye, and at its faintest, you need a telescope to spot it. The changes of a long period variable star are also much more variable than the changes of a Cepheid. The brightest magnitude that a particular long period variable star reaches can be quite different from one period to the next. Such changes are easily observed, and they constitute basic scientific information. You can help in these and other variable star studies, as I describe in the last section of this chapter.

Explosive neighbors: Flare stars

Flare stars are little red dwarfs that suffer big explosions, like ultrapowerful solar flares. You can't see most solar flares without the aid of special colored filters because the light of the flare is just a tiny fraction of the total light of the Sun. Only the rare, very large "white light" flares are visible on the Sun without a special filter. (But you still need to use one of the techniques for safe solar viewing that I describe in Chapter 10.) However, the explosions on flare stars are so bright that the magnitude of the star changes detectably. You're looking at the star through a telescope, and suddenly it shines brighter. Not all red dwarfs have these frequent explosions. Proxima Centauri, the nearest star beyond our Sun, is a flare star.

Nice to nova: Exploding stars

The explosions of novas and supernovas are so large that I can't lump them in with the flare stars; they're enormously more powerful and have much greater effects.

Novas

A nova explodes through a build-up process on a white dwarf in a binary system, much like the Type Ia supernova explosions that I describe earlier in this chapter.

But in a supernova, the white dwarf is shattered; in a nova, the white dwarf isn't destroyed. The star just blows its top, and then it settles down, sucking more gas off its companion and onto its surface layer. The powerful gravity of the white dwarf compresses and heats this layer, and after centuries or millennia, off it goes again. At least, that's the theory. No scientist has been around long enough to see an ordinary or *classical nova* explode twice. But similar binary systems exist in which the explosions aren't quite as powerful as in a classical nova, but in which the explosions recur frequently enough that amateur astronomers are always monitoring them, ready to announce the discovery of a new explosion and alert professionals to study it. These objects have various names, including *dwarf nova* and *AM Herculis systems.*

Classical novas, dwarf novas, and similar objects are known collectively as *cataclysmic variables.*

A nova bright enough to see with the naked eye occurs about once a decade, give or take. I studied one in Hercules for my doctoral thesis in 1963. If it hadn't exploded at just the right time, I may have waited ten years for a thesis topic. Most recently, a bright nova in Scorpius dazzled astronomers in 2007.

Supernovas

Supernovas throw off nebulae, called *supernova remnants,* at high speeds (see Figure 11-5). The nebula at first consists of the material that made up the shattered star, minus any remaining central object, be it neutron star or black hole (see the section "Closing time: Coming up on the tail end of stellar evolution," earlier in this chapter). But as it expands into space, the nebula sweeps up interstellar gas like a snowplow that accumulates snow. After a few thousand years, the supernova remnant is mostly swept-up gas rather than supernova debris.

Figure 11-5:
A portion of the Cygnus Loop, a supernova remnant.

Courtesy of NASA

Supernovas are incredibly bright and rather rare. Astronomers estimate that, in a galaxy like the Milky Way, a supernova occurs every 25 to 100 years, but we haven't witnessed a supernova in our home galaxy since Kepler's Star in 1604, before the invention of the telescope. Others may have occurred, obscured from view by dust clouds in the Galaxy. A huge southern star known as Eta Carinae looks like it may be on the verge of going supernova in the Milky Way, but in astronomers' parlance, that means it may explode at any time within the next million years.

Only one supernova has been visible with the naked eye since 1604. It was Supernova 1987A, located in our neighbor galaxy, the Large Magellanic Cloud, or LMC (which I describe in Chapter 12). The supernova was too far south to be seen from the continental United States, but I wasn't going to miss a celestial event of such great rarity, so I flew to Chile to see it. The Chilean astronomers gave me a warm reception.

Hypernovas

Hypernovas are especially bright supernovas that seem to produce at least some of the gamma ray bursts that flash in the sky from time to time. The bursts are extremely powerful blasts of high-energy radiation emitted in beams like the beams from searchlights. NASA launched the Swift satellite in November 2004 to discover more about them. When Swift detects a burst coming from a certain direction, it, well, swiftly notifies observatories on the ground to focus on that part of the sky. Hypernovas are much rarer than other supernovas, and no hypernova has been seen in our own Galaxy.

If you want to know more about Swift and its discoveries, go to NASA's Swift website at www.nasa.gov/mission_pages/swift/main/index.html and also check out the Swift Education and Public Outreach site at http://swift.sonoma.edu. If you've got an iPad or an iPhone, just open iTunes and download the free Swift Explorer App from Pennsylvania State University — it has some cool capabilities. You can even set it to send a message to your phone each time Swift detects a gamma ray burst.

Stellar hide-and-seek: Eclipsing binary stars

Eclipsing binary stars are binary systems whose true brightness doesn't change (unless one of the two stars happens to be a pulsating star, flare star, or other intrinsic variable), even though they look like variable stars to us. The *orbital plane* of the system — the plane that contains the orbits of the two stars — is oriented so that it contains our line of sight to the binary system. So on every orbit, one star eclipses the other, as seen from Earth, and the star's brightness dips during the eclipse. (Of course, the situation reverses halfway through the orbital period, when the eclipsed star now does the eclipsing.)

If the two stars in a binary system have orbital periods of four days, then every four days, the more massive star in the system, usually called *A,* passes exactly in front of the other star in the view from Earth. This pass blocks all or most of the light from star *B* from reaching Earth (depending on its size compared to *A* — sometimes the less massive star is larger than its heavy companion), so the binary looks fainter. Astronomers call this event a *stellar eclipse.* Two days after the eclipse, star *B* passes in front of star *A,* creating another eclipse.

In the earlier section "Binary stars and the Doppler effect," I mention how astronomers use the orbital velocities to figure out the relative masses of the stars. Well, they can also use the velocities to find the diameters of stars. Scientists take spectra and discover how fast the stars orbit by using the Doppler effect, and they measure the durations of the eclipses in eclipsing binaries. A stellar eclipse of star *B* begins when the leading edge of *A* starts to pass in front of it. The eclipse ends when the following edge of *A* finishes passing in front of *B.* So the orbital speed multiplied by the duration of the eclipse tells scientists how big star *A* is.

In all these methods, the fine details are a little more complicated, but you can easily understand the principles of stellar investigation.

The most famous eclipsing binary is Beta Persei, also known as Algol, the Demon Star. You won't have a devil of a time observing Algol's eclipses in the Northern Hemisphere — Algol is a bright star well placed for observation in the northern sky in the fall. You can watch its eclipses without a telescope or even binoculars. Every 2 days and 21 hours, Algol's brightness dims by a little more than one magnitude — more than a factor of 2.5 — for about two hours. But you need to know *when* to look for an eclipse. You don't want to stand around in the backyard for almost three days. The neighbors will talk. Check out the pages of *Sky & Telescope* that list information for observers. Usually, you find a paragraph called "Minima of Algol," which lists the dates and times when eclipses will occur during a period of a month or two. (If you don't see a list in the current issue, it means that Algol is too close to the Sun in the sky for observation that month.)

Minima are the times when variable stars, extrinsic or intrinsic, reach the lowest brightness levels of their ongoing cycles. *Maxima* are the times when the stars shine brightest.

Hog the starlight: Microlensing events

Sometimes a faraway star passes precisely in front of an even farther star. The two stars are unrelated and may be thousands of light-years from each other, but the gravity of the nearer star bends the paths taken by light rays from the farther star in such a way that the distant star appears much brighter from Earth for a few days or weeks. This effect is predicted from

Einstein's Theory of General Relativity, and astronomers regularly detect it. When the gravity of a huge object like a galaxy bends light, astronomers call the process *gravitational lensing*. When the gravity of a body as small as a star bends light, it's called *microlensing*.

You may be thinking how unlikely it is that two unrelated stars would line up perfectly with Earth, and you would be right! Congratulations on that thought. To detect such a rare event on a regular basis, astronomers use telescopic electronic cameras that can record hundreds of thousands to millions of stars at a time. With that many stars under observation, a foreground star passes in front of a distant star every so often, even though astronomers don't know beforehand which two stars are involved.

The trick is to point the electronic camera at a region where you can see vast numbers of stars simultaneously in the field of view. Such regions include the Large Magellanic Cloud, a nearby satellite galaxy of the Milky Way (see Chapter 12), and the central bulge of the Milky Way itself, where a whole mess of stars hang out.

Your Stellar Neighbors

When you look at Alpha Centauri with the naked eye, you see one bright star. When you check it out with a telescope, you see two bright stars close together in the field of view. The two stars form a binary system. But a third star, Proxima Centauri, makes it a triple system. You don't see Proxima in the field of view with the two bright stars because it's in a huge orbit around them and, as seen from Earth, is more than 2° away, or more than four times the apparent diameter of the full Moon. (And Proxima, a faint red dwarf star, isn't visible to the naked eye. You've already met it in the section "Main sequence stars: Enjoying a long adulthood," earlier in this chapter.).

For a complete look at the triple system, check out the following list:

- ✔ **Alpha Centauri (also called Rigil Kentaurus):** This star is a bright, G-type star in the southern constellation Centaurus (see Figure 11-6). It's a main sequence dwarf with about the same color as the Sun but is somewhat brighter.

- ✔ **Alpha Centauri B:** Alpha Centauri's orange companion is a slightly smaller and cooler main sequence dwarf.

- ✔ **Alpha Centauri C:** Our nearest stellar neighbor beyond the Sun, the little red dwarf and flare star, is also called Proxima Centauri.

Figure 11-6:
Alpha
Centauri is
a triple-star
system
in the far
southern
sky.

The Alpha Centauri system is about 4.4 light-years from Earth, with Proxima on the near side, at 4.2 light-years. The system is in the far southern sky, so you need to be in the Southern Hemisphere, or at least very far south in the Northern Hemisphere, to view it.

Sirius, at a distance of 8.5 light-years, is the brightest star in the night sky. Its official name is Alpha Canis Majoris, in Canis Major (the Great Dog; see Figure 11-7). Slightly south of the Celestial Equator, Sirius is easily visible from most inhabited places on Earth. It's a white, A-type, main sequence star that shines bright enough to make folks ask, "What's that big star?"

Figure 11-7:
Sirius is top
dog in Canis
Major.

Like most stars other than the Sun, Sirius has a companion star: Sirius B, a white dwarf. Sirius is known as the Dog Star, and when the American telescope maker Alvan Clark discovered its tiny companion, Sirius B, in 1862, naturally someone nicknamed it "the Pup."

One legend and some written records that are open to differing interpretations suggest that Sirius was a red star a few thousand years ago. Despite much effort, astrophysicists have been unable to explain this color in terms of known physical processes, so naturally, they say the story isn't true.

Vega is Alpha Lyrae, the brightest star in the constellation Lyra (the Lyre). It appears high in the sky at temperate northern latitudes (such as the U.S. mainland) on summer evenings and is an object that most amateur astronomers know like the back of their hands. Located about 26 light-years from Earth, Vega is an A-type main sequence star like Sirius. Vega is a brilliant white sparkler and one of the most conspicuous stars in the sky.

Betelgeuse isn't really in the solar neighborhood; it's a red supergiant star of spectral type M, about 640 light-years from Earth. But everybody likes its name, which many pronounce "Beetle Juice" (which is as good a way to say it as any), and observers enjoy its deep red color. It is, after all, a red supergiant, more than 20,000 times brighter than the Sun. Although Betelgeuse is Alpha Orionis, the brightest star in Orion is actually Rigel (Beta Orionis).

How to Help Scientists by Observing the Stars

Thousands of stars are under watch because they vary in brightness or exhibit some other special characteristic. Professional astronomers can't keep up with them all, and that's where you come in. You can monitor some stars with your own eyes, with binoculars, or with a telescope.

You need to be able to recognize the stars and judge their magnitudes. The brightness of many stars changes so significantly — by a factor of two, ten, or even hundreds — that eye estimates are sufficiently accurate to keep track of them. The trick is to use a *comparison chart*, a map of the sky that shows the position of the variable star and the positions and magnitudes of *comparison stars*. A comparison star has a known brightness that doesn't vary (or so we hope).

The American Association of Variable Star Observers (AAVSO) offers a wealth of information explaining how to observe variable stars. Its website is www.aavso.org.

AAVSO encourages beginners and expert amateur observers alike. You can download its *Manual for Visible Observing of Variable Stars* in your choice of English or various other languages. (Amateur astronomers around the world contribute their observations of variable stars to AAVSO's research.)

Check out AAVSO's Variable Star Plotter (VSP) on its website. You can enter the name or number of a variable star, and VSP will create a sky chart for the star that you can download for use with your telescope. After you've read the manual and practiced judging star magnitudes, you'll be all set to observe the variable star and send your findings to AAVSO.

Star Studies to Aid with Your Brain and Computer

If you live in a place where weather conditions and/or urban lighting are not conducive to astronomical observations on a frequent basis, you can still help astronomers by practicing *citizen science:* You can examine telescopic data from orbiting spacecraft or from ground-based professional observatories, in accordance with online instructions. All you need to do is sign up on a project website, study the project tutorial, and begin examining the data.

Thousands of interested persons enroll in these projects, so although any one individual may not have the best scientific judgment, the reports from many participants average out. The citizen scientist findings generally alert professional astronomers to interesting and possibly scientifically important objects or phenomena in the (literally) astronomical databases that no one expert can scan by himself.

Here are two good citizen science projects to consider:

✔ **The Milky Way Project (**www.milkywayproject.org/tutorial**):** This research project seeks new information on how stars form.

Sponsored by the Adler Planetarium in Chicago, Illinois, and the Zooniverse Citizen Science program, this project uses images from NASA's Spitzer Space Telescope. When you participate, you use the computer tools provided on the MWP site to draw circles around so-called *green knots* in the Milky Way (examples are given so you know what to look for). The knots may be places where star formation is underway. You also hunt for small, previously unknown star clusters and circle them. Professional astronomers use the information you submit to benefit ongoing research on the stars. (I describe the Milky Way and its star clusters in Chapter 12.)

✔ **The Search for Exploding Stars (Supernovae)** (http://supernova.galaxyzoo.org): You can hunt supernovae from the comfort of your home by joining The Search for Exploding Stars (Supernovae). When the Exploding Stars project is operating regularly, you receive telescopic images from an automated camera at Palomar Observatory in California and search them for new supernovae. (I said "when" because, as of early May 2012, volunteers had exhausted the available Palomar data, and newcomers to the website were being directed temporarily to other worthwhile citizen science projects.)

If you do join Exploding Stars, you'll find out how to identify and report a supernova. Based on reports from you and other volunteers, professional astronomers point their powerful telescopes at the exploding stars and gather fresh astrophysical information.

I think Exploding Stars beats *Angry Birds* any day of the week, plus it's free, and no animals are harmed (unless they're on a planet close to a supernova).

Chapter 12

Galaxies: The Milky Way and Beyond

*O*ur solar system is a tiny part of the Milky Way galaxy, a great system of hundreds of billions of stars, thousands of nebulae, and hundreds of star clusters. The Milky Way, in turn, is one of the largest components of the Local Group of Galaxies. Beyond the Local Group is the Virgo Cluster, the nearest large cluster of galaxies — about 54 million light-years from Earth. As scientists peer out into the universe over much greater distances, they see *superclusters,* immense systems that contain many individual clusters of galaxies. So far, we haven't found superclusters of superclusters, but *Great Walls,* which are immensely long superclusters, do exist. And much of the universe seems to consist of gigantic cosmic voids, which contain relatively few detectable galaxies.

This chapter introduces you to the Milky Way and its most important parts, and takes you farther into the universe to meet other types of galaxies.

Unwrapping the Milky Way

The Milky Way, also called "the Galaxy," is a lot bigger than the candy bar, if not as sweet. But it does have a creamy-looking center. The Milky Way is seen from Earth as the wide band of diffuse light that you can view best on clear summer and winter nights from a dark location.

Uncloaking the murky Milky Way

In the past, stargazers viewed the Milky Way with ease, but now many people can't see it or don't know that it exists because they live in or near cities with so many lights that the sky is bright with light pollution rather than dark as nature intended.

The solution? Flee the light pollution, at least once in awhile. Go out to the mountains or the shore on your vacation or on a weekend and check out a darker sky than you have at home.

The light of the full Moon interferes with seeing the Milky Way, too, so plan your jaunt around the time of the new Moon, when there is little or no moonlight in the sky. The Milky Way is most prominent in the sky during summer and winter, and least visible in spring and fall. (For tips on light pollution and dark-sky sites for stargazing, visit the International Dark-Sky Association website, at www.darksky.org.)

A stream of milk through the universe was as good as any explanation for the Milky Way until 1610, when Galileo first observed it with a telescope. He found that the Milky Way is nothing to lap about: It consists of a huge number of stars (today estimated at about 300 billion), most of them so faint or so far away that they blend into one large fuzzy region on the sky. Most of the stars in the Milky Way are invisible to the eye, but as a group, they shine. Obviously, the telescope was a definite improvement for studying the Milky Way (and almost everything else in astronomy)!

Galaxies are the basic building blocks of the universe, and the Milky Way is a good-size block. It contains almost everything in the sky that you can see with the naked eye — from Earth and our solar system to the stars of the solar neighborhood, the visible stars in the constellations, and all the stars that blend together to make that milky stream in the night sky — and plenty of objects and matter that you can't see. It also contains nearly every nebula you can see without telescopic aid — and plenty more, to boot.

Now that's a tall glass of milk! Besides loose stars, the Milky Way holds hundreds of star clusters, such as the Pleiades and Hyades in the constellation Taurus and, for lucky viewers in Australia, South America, and other points far south, the Jewel Box in Crux, the (Southern) Cross, and the magnificent star cluster Omega Centauri.

How and when did the Milky Way form?

The Milky Way is almost as old as the universe, and it certainly goes back at least as far as the 13 billion years scientists estimate to be the age of some of its oldest stars. Long ago, gravity caused a gigantic cloud of primordial gas to fall together and condense. As little clumps inside the cloud collapsed even faster than the cloud as a whole, stars formed. Although the big cloud must have turned very slowly at first, it would've rotated faster and faster as it became smaller and flattened into the present-day spiral disk structure. And before you knew it, *voilà, la voie lactee* (French for "there it is, the Milky Way"). Actually, its formation isn't quite that simple because the Milky Way is a glutton — it has swallowed smaller neighbor galaxies for eons, adding their stars to its own collection. It continues its feast today. What a menace!

You just read my favorite Milky Way theory, bar none. If you have a better one, become an astronomer yourself and write your own book someday — in science, theories make the world go 'round, and maybe even the Galaxy.

What shape is the Milky Way?

The Milky Way is the shape it is and the size it is because, in the universe, gravity rules. The Milky Way is a spiral galaxy, consisting of a pizza-shape formation of billions of stars (the *galactic disk,* about 100,000 light-years in diameter) that contains the spiral arms (see Figure 12-1). The arms are shaped roughly like the streams of water from a rotating lawn sprinkler and contain many bright, young blue and white stars and gas clouds. Groups of young, hot stars (called *associations*) dot the spiral arms in the galactic disk like pepperoni slices on a pizza. Bright and dark nebulae seem to mushroom all around the arms, along with huge molecular clouds, such as Monoceros R2 (its location is marked in Figure 12-1), where most of the gas is cool and dim. Between the arms are the *interarm regions* (not all astronomical terms are as catchy as Barnacle Bill, the name of a rock on Mars, or the Red Rectangle, a nebula shaped like an hourglass — go figure).

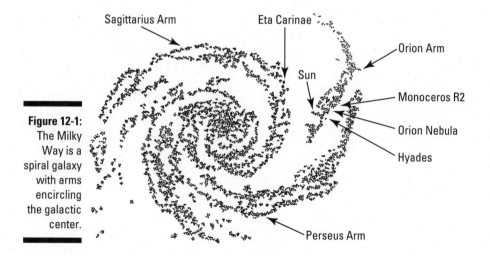

Figure 12-1:
The Milky Way is a spiral galaxy with arms encircling the galactic center.

At the center of our galaxy is a place called (you guessed it!) the *galactic center*. Centered on the center is the galactic bulge, which puts a sumo wrestler to shame. The *galactic bulge* is a roughly spherical formation of millions of mostly orange and red stars, sitting like a great meatball at the center of the galactic disk and extending far above and below it. But at least some of the bulge stars are arranged in an elongated formation, more sausage-shaped than meatball-shaped. Astronomers call the sausage a bar. When a spiral galaxy has a prominent bar, it's called a *barred spiral* (I discuss barred spirals later in this chapter), but the bar in the Milky Way isn't very evident.

At the center of the galactic bulge is Sagittarius A*, a supermassive black hole. Figure 12-1 presents a model of the Milky Way with its toppings and ingredients. (It's a close-up of the galactic disk, with the galactic bulge omitted for clarity.)

Any way you slice it, the Milky Way is a terrific galaxy.

The flat imaginary surface or midplane of the galactic disk is called the *galactic plane,* and the circle that represents its intersection with the sky, as visible from Earth, is called the *galactic equator.*

Sometimes astronomers list the position of an object in galactic coordinates rather than in right ascension and declination (coordinates that I define in Chapter 1). The galactic coordinates are *Galactic Latitude,* which is measured in degrees north or south of the galactic equator, and *Galactic Longitude,* which is measured in degrees along the galactic equator.

Galactic Longitude starts at the direction to the galactic center, which is 0° longitude. (Actually, the zero point of Galactic Longitude is slightly off the galactic center because scientists placed it where the galactic center was thought to be in 1959; we know better now.) Galactic Longitude proceeds along the galactic equator from the constellation Sagittarius into Aquila, Cygnus, and Cassiopeia. It goes on through Auriga, Canis Major, Carina, and Centaurus, and all the way to 360° longitude, back at the galactic center. When you look with binoculars at the constellations I just named, you see more stars, star clusters, and nebulae than elsewhere in the sky. The "plane truth" is that the constellations that the galactic plane intersects are among the finest sights in the sky.

You can find panoramic maps of the Milky Way's galactic plane — as recorded with radio telescopes, X-ray and gamma-ray observatory satellites, and visible light (or "optical") telescopes on the ground — at NASA's MultiWavelength Milky Way website, at http://mwmw.gsfc.nasa.gov.

Seeing beyond the Milky Way

The three objects that are beyond the Milky Way but readily visible with the naked eye are the Large and Small Magellanic Clouds (two nearby galaxies visible from the Southern Hemisphere) and the Andromeda Galaxy. Some people blessed with excellent vision (and many others who want to impress their friends) say that they can see the Triangulum Galaxy, too. Both the Andromeda and Triangulum galaxies are about 2 million light-years from Earth, but Andromeda is bigger and brighter.

I count the Large Magellanic Cloud as one object, but it actually contains a huge bright nebula, the Tarantula, which you can make out with the naked eye. (Don't worry, it won't bite.) For a few months in 1987, you could see a bright supernova in the Large Cloud, Supernova 1987A. It was the first supernova visible to the naked eye since Kepler's Star in 1604, although it wasn't in our own galaxy like Kepler's Star. Unlike Kepler's Star, Supernova 1987A wasn't visible from Europe or the continental United States, but was readily seen from places like Australia, Chile, and South Africa.

Where can you find the Milky Way?

The Milky Way isn't a certain distance from Earth; it contains our planet. The galactic center is about 27,000 light-years from Earth. Recent measurements with a radio telescope named the *Very Long Baseline Array* show that the solar system takes about 226 million years to make one orbit around the galactic center. The measurement cleared up a discrepancy: Scientists were unsure whether this interval, the *galactic year,* was 200 million years or 250 million years. Now they have an accurate number.

The outskirts of the galactic disk — or *galactic rim,* as science fiction fans know it — is, at its closest part to Earth, about an equal distance in the opposite direction from the galactic center. The disk of the Milky Way is pretty much identical to the milky band of light in the sky.

The Milky Way is about 169,000 light-years from a galaxy called the Large Magellanic Cloud, about 2.6 million light-years from the Andromeda Galaxy, and about 54 million light-years from the nearest big cluster of galaxies, the Virgo Cluster. It also falls smack dab in the middle of a little cluster of galaxies (sizes are relative here), the Local Group.

Star Clusters: Meeting Galactic Associates

Star clusters are bunches of stars located in and around a galaxy. They haven't associated by chance (even though one type of star cluster is called an *association*); they're groups of stars that formed together from a common cloud and, in most cases, are held together by gravity. The three main types of star clusters are *open clusters, globular clusters,* and *OB associations*.

For superb images of star clusters, consult the Anglo-Australian Observatory at www.aao.gov.au/images, or treat yourself to a coffee-table book — *The Invisible Universe,* by David Malin (Bulfinch Press, 1999) — that contains a collection of the greatest photographs from the Anglo-Australian Observatory. You can pick up the book in French, German, Italian, and Japanese editions because everyone loves these great pictures. You can also check out two photos of star clusters in the color section of this book.

A loose fit: Open clusters

Open clusters contain dozens to thousands of stars, have no particular shape, and are located in the disk of the Milky Way. Typical open clusters span about 30 light-years. They aren't highly concentrated (if at all) toward their centers, unlike globular clusters (see the next section), and open clusters typically are much younger. Open clusters are great for viewing with small telescopes and binoculars, and you can see some with the naked eye. You can find them marked on most good star atlases, such as *Sky & Telescope's Pocket Sky Atlas,* by Roger W. Sinnott (New Track Media, 2006). In this atlas, open clusters are marked by yellow disks with dotted edges, and the disk sizes are proportional to the apparent sizes of the clusters as seen from Earth. (The atlas also shows globular clusters, which I describe in the next section.)

The most famous and easily seen open clusters in the northern sky are as follows:

- ✔ **The Pleiades (also known as the Seven Sisters):** Located in the northwest corner of Taurus, the Bull, the Pleiades looks to the naked eye like a tiny dipper. You can compare your eyesight with a friend's by seeing how many stars you can count in the Pleiades, which is M45, the 45th object in the *Messier Catalog* (see Chapter 1). Gaze through binoculars and see how many more you can find. The brightest star in the Pleiades is Eta Tauri (3rd magnitude), also called Alcyone. (See Chapter 1 for an explanation of magnitudes.)

- ✔ **The Hyades:** Also located in Taurus and another great sight for the naked eye, the Hyades includes most of the stars that make up the V shape in the head of Taurus. You can't miss this cluster because the V features the bright red giant Aldebaran (1st magnitude), which is Alpha Tauri (see Figure 12-2). Aldebaran actually is far beyond the Hyades cluster, but it appears in the same direction from Earth.

The Hyades looks much bigger than the Pleiades because it's about 150 light-years from Earth versus about 400 light-years for the Pleiades.

✔ **The Double Cluster:** Located in Perseus, the Hero, the Double Cluster is a beautiful sight through binoculars and especially through a small telescope. Its two clusters are NGC 869 and 884, each roughly 7,000 light-years from Earth. NGC stands for *New General Catalogue,* which was new when it first appeared in 1888, and it didn't list any generals (or captains or colonels, for that matter).

✔ **The Beehive (also called Praesepe):** The Beehive (Messier 44) is the main attraction of Cancer, the Crab, a constellation composed of dim stars. It looks like a nice fuzzy patch with the naked eye and like a swarm of many stars when seen through binoculars. This cluster is about 600 light-years from Earth.

Figure 12-2: Taurus contains the red giant Aldebaran (Alpha Tauri).

Viewers in the Southern Hemisphere can enjoy these open clusters:

✔ **NGC 6231:** In Scorpius, the Scorpion, NGC 6231 is a southern-sky object, but you can readily see it from much of the United States in the evening during the summer. You need to be in a dark place with an unobstructed southern horizon. The observer Robert Burnham, Jr., described it as resembling "a handful of diamonds on black velvet."

✔ **NGC 4755 (also called the Jewel Box):** Located in Crux, the Cross, the Jewel Box includes the bright star Kappa Crucis. Crux, popularly known as the Southern Cross, is a perennial favorite of viewers in the Southern Hemisphere. If you take a cruise through the South Seas, insist that the ship have an astronomy lecturer on board. (I will probably be available.) He or she can point out the Southern Cross; with binoculars, you can enjoy the fine sight of the Jewel Box.

A tight squeeze: Globular clusters

Globular clusters are the retirement homes of the Milky Way Galaxy. They appear about as ancient as the galaxy itself (some experts think that the globular clusters were the first objects to form in the Milky Way), so they contain ancient stars, including many red giants and white dwarfs (see Chapter 11). The stars you can see in a globular cluster with your telescope are mostly red giants. With bigger telescopes, you can see orange and red main sequence dwarf stars. Only the Hubble Space Telescope and other very powerful instruments can pick out many of the much fainter white dwarfs in a globular cluster.

A typical globular star cluster contains a hundred thousand to a million or more stars, all packed in a ball (hence the term *globular*) that measures just 60 to 100 light-years in diameter. The closer the stars are to the center, the more tightly they're packed in (see Figure 12-3). Its high degree of concentration and its great number of stars distinguish a globular cluster from an open cluster.

Another key difference is that open clusters are distributed across the galactic disk in a great flat pattern, but globular clusters are arranged spherically around the center of the Milky Way. Most globular clusters are concentrated toward the galactic center, but many of the globulars you can easily see are well above or below the galactic plane.

Here are the best globular clusters for viewing in the northern sky:

- **Messier 13:** This globular cluster makes its home in Hercules, representing the mythical character of the same name.
- **Messier 15:** You can find this globular cluster in Pegasus, the Winged Horse.

Figure 12-3:
Globular cluster G1 in the Andromeda Galaxy.

Courtesy of NASA

You can spot both M13 and M15 with the naked eye under suitably dark sky conditions, but you may need to reassure yourself with binoculars or a small telescope, which show the clusters as fuzzy spots larger than stars. Use a star chart (such as *Sky & Telescope's Pocket Sky Atlas,* which I mentioned in the preceding section) to locate the clusters.

Observers in the Northern Hemisphere are cheated out of the best globular star clusters because by far the two biggest and brightest shine in the deep southern sky:

- ✔ **Omega Centauri:** Located in Centaurus, the Centaur
- ✔ **47 Tucanae:** In Tucana, the Toucan

These clusters are spectacular sights through small binoculars and just about worth the trip to South America, South Africa, Australia, or other places where you can readily see them. Bring a friend because two can share a view of the Toucan.

Be sure to check out the photo of a globular cluster in the color section.

Fun while it lasted: OB associations

OB associations are loose stellar groupings with dozens of stars of spectral types O and B (the hottest types of main sequence stars) and sometimes fainter, cooler stars (see Chapter 11 for more about spectral types). Unlike with open clusters and globular clusters, gravity doesn't hold together OB associations; over time, the stars move away from each other, dissolving the association like a limited partnership that is breaking up. OB associations are located close to the galactic plane.

Many of the bright young stars in the constellation Orion are members of the Orion OB association. (See Chapter 3 for more about Orion.)

Taking a Shine to Nebulae

A nebula is a cloud of gas and dust in space ("dust" meaning microscopic solid particles, which may be made of silicate rock, carbon, ice, or various combinations of those substances; and "gas" meaning hydrogen, helium, oxygen, nitrogen, and more, but mostly hydrogen). As I note in Chapter 11, some nebulae play an important role in star formation; others form from stars gasping on their deathbeds. Between the cradle and the grave, nebulae come in a number of varieties. (Check out a photo of a nebula in the color section of this book.)

Here are a few of the most familiar nebulae:

- ✔ **H II regions** are nebulae in which hydrogen is ionized, meaning that the hydrogen loses its electron. (A hydrogen atom has one proton and one electron.) The gas in an H II region is hot, ionized, and glowing, due to the effects of ultraviolet radiation from nearby O or B stars. All the large, bright nebulae that you can see through binoculars are H II regions. (*H II* refers to the ionized state of the hydrogen in the nebula.)

- ✔ **Dark nebulae,** also known as H I regions, are the dust bunnies of the Milky Way, consisting of clouds of gas and dust that don't shine. Their hydrogen is neutral, meaning that the hydrogen atoms haven't lost their electrons. The term *H I region* refers to the neutral (non-ionized) state of the hydrogen.

- ✔ **Reflection nebulae** consist of dust and cool, neutral hydrogen. They shine by the reflected light of nearby stars. Without the nearby stars, they would be dark nebulae.

    Sometimes a new reflection nebula appears suddenly, and you may discover it, as amateur astronomer Jay McNeil did. In January 2004, he found a new reflection nebula in the constellation Orion with a 3-inch refractor in his backyard, and professionals now call it McNeil's Nebula. But don't hold your breath; this type of discovery is rare.

- ✔ **Giant molecular clouds** are the largest objects in the Milky Way, but they're cold and dark, and scientists would've gazed right by them if not for the data gathered by radio telescopes, which can detect emissions of faint radio waves from molecules in these clouds, such as carbon monoxide (CO). Like all other nebulae, giant molecular clouds are mostly made of hydrogen, but scientists often study them by means of their trace gases, such as CO. The hydrogen in giant clouds is molecular, with the designation H_2, which means that each molecule consists of two neutral hydrogen atoms.

    One of the most exciting nebular discoveries of the 20th century was that bright H II regions, such as the Orion Nebula, are hot spots on the peripheries of unseen giant molecular clouds. For centuries, people could see the Orion Nebula but didn't know that it's no more than a bright pimple on a huge invisible object, the Orion Molecular Cloud. According to current ideas, new stars are born in molecular clouds, and when they get hot enough, they ionize their immediate surroundings, turning them into H II regions. The part of a molecular cloud where the dust is thick enough to cut off the light of many or most of the stars behind the cloud, as visible from Earth, is called a dark nebula.

H II regions, dark nebulae, giant molecular clouds, and many of the reflection nebulae are located in or near the Milky Way's galactic disk.

Two other interesting types of nebulae are planetary nebulae and supernova remnants, which I cover briefly in the following sections (and in Chapter 11).

Picking out planetary nebulae

Planetary nebulae are produced by old stars that started out resembling the Sun but then expelled their outer atmospheric layers. (The Sun will do the same thing in the far future (see Chapter 10).) The expelled gases make up the nebulae, which are ionized and made to glow by ultraviolet light from the hot little stars at their centers. The tiny stars are all that remain of the former suns. Planetary nebulae expand into space and fade as they grow larger. They can be well off the galactic plane, unlike H II regions. (See a planetary nebula in the color section of this book.)

For decades, astronomers believed that many or most planetary nebulae were roughly spherical. But now we know that most of these nebulae are *bipolar,* meaning that they consist of two round lobes projecting from opposite sides of the central star. The planetary nebulae that look spherical, such as the Ring Nebula in the constellation Lyra (see Figure 12-4), are bipolar, too, but the axis down the center of the lobes happens to point toward Earth (and so, like a dumbbell viewed end on, they look circular). Astronomers took many years to figure this out, so maybe we're dumbbells, too.

Figure 12-4:
The Ring Nebula in Lyra is bipolar but appears spherical from Earth.

Courtesy of NASA

Correcting a galactic goof-up

Until the 1950s, astronomers used the term *nebula* to refer to a galaxy because, until the 1920s, they thought galaxies other than the Milky Way were nebulae in the Milky Way. In other words, astronomers believed that there was only one galaxy: the Milky Way.

It took a few dozen years for the change in understanding to prevail in the language of astronomy. Therefore, the authors of astronomy books have only recently stopped referring to the Andromeda Galaxy as the Andromeda Nebula.

Edwin P. Hubble, for whom the Hubble telescope is named, wrote the famous book *The Realm of the Nebulae*. The book was all about galaxies, not nebulae as astronomers use the term now. Among his achievements, Hubble proved that the Andromeda Nebula is a galaxy full of stars, not a big cloud of gas. A former boxer, he fought in World War I, smoked a pipe, and supposedly bullied other astronomers at Mount Wilson Observatory, but his discoveries are no bull.

Curious point: *Protoplanetary nebulae* are much studied by astrophysicists, but there are actually two kinds, and they have nothing to do with each other. One type of protoplanetary nebula is the early stage of a planetary nebula — a phase in the death of a star (not to be confused with the *Star Wars* Death Star). The other type of protoplanetary nebula is the birth cloud of a solar system's star and its planets. Yes, astronomers use the same term to refer to two completely different kinds of objects, but nobody's perfect.

Breezing through supernova remnants

Supernova remnants begin as material ejected from massive stellar explosions. A young supernova remnant consists almost exclusively of the shattered remains of the exploded star that expelled it. But as the gas moves outward through interstellar space, it resembles a rolling stone that *does* gather moss. The expanding remnant creates a snowplow effect as it pushes along (see Chapter 11) and accumulates the thin gas of interstellar space. After some thousands of years, the remnant is overwhelmingly composed of this "plowed up" interstellar gas, and the remains of the exploded star are mere traces. Supernova remnants are located along or near the galactic plane of the Milky Way.

Enjoying Earth's best nebular views

Nebulae are among the most beautiful sights through small telescopes. You need a good star chart, like those found in *Sky & Telescope's Pocket Sky Atlas*, and you want to start with an easy target, like the Orion Nebula, which you glimpse by eye and binoculars before homing in on it with your telescope. For H II regions like the Orion Nebula, a telescope with a low f/number, such as the

Orion ShortTube 80 Equatorial Refractor, may work the best (see Chapter 4, where I discuss using the scope for hunting comets, for more info on this particular tool). For smaller nebulae like the Ring Nebula, which I describe in the following list, the Meade ETX-90 telescope (see Chapter 3) is a good choice for a beginner. The Meade has a computerized control system that points the telescope accurately at objects that you can't see with the naked eye.

The following are some of the best, brightest (or, for dark nebulae, darkest), and most beautiful nebulae you can see from northern latitudes, including some southern-sky objects that aren't located very far south of the celestial equator:

✔ **The Orion Nebula,** Messier 42 (see Chapter 1), in Orion, the Hunter

You can easily see the Orion Nebula, an H II region, with the naked eye as a fuzzy spot in the sword of Orion. It looks fine through binoculars and spectacular through a small telescope. The telescope also shows the Trapezium, a bright quadruple star (see Chapter 11) in the nebula.

✔ **The Ring Nebula,** Messier 57, in Lyra

The Ring Nebula is a planetary nebula high in the sky at northern temperate latitudes on a summer evening. As with all planetary nebulae, you need to use a star chart to find it with your telescope, unless you have a computer-assisted telescope such as the Meade ETX-90 (see Chapter 3), which points right to the nebula at your command.

✔ **The Dumbbell Nebula,** Messier 27, in Vulpecula, the Little Fox

The Dumbbell Nebula, along with the Ring Nebula, is among the easiest planetary nebulae to spot with a small telescope. The best time for observation is the summer and fall.

✔ **The Crab Nebula,** Messier 1, in Taurus

The Crab Nebula is the remains of a supernova that exploded in the year 1054, as seen from Earth. It appears as a fuzzy spot through a small telescope, but a big telescope shows two stars near its center. One star isn't associated with the Crab; it just appears along the same line of sight. The other star is the pulsar (see Chapter 11) that remains from the supernova explosion. It spins 30 times per second, and one or the other of its two lighthouse beams sweeps across Earth every $\frac{1}{60}$ of a second, with the same frequency as your 60-cycle, alternating-current household electricity. (I think that's a "powerful" analogy.)

✔ **The North American Nebula,** NGC 7000, in Cygnus, the Swan

The North American Nebula (the name comes from its shape) is a faint but large H II region that you can see with the naked eye on a summer evening at a dark location with no moonlight. To glimpse it, use averted vision — look out of the corner of your eye.

✔ **The Northern Coal Sack,** in Cygnus, the Swan

The Northern Coal Sack is a dark nebula near Deneb, which is Alpha Cygni, the brightest star in Cygnus. You can recognize the Coal Sack by eye as a dark blotch against the brighter background of the Milky Way.

Don't skip the following nebulae located at moderate southern declinations. You can view them from many places in the Northern Hemisphere and from anywhere in the Southern Hemisphere:

- ✔ **The Lagoon Nebula,** Messier 8, in Sagittarius, the Archer

- ✔ **The Trifid Nebula,** Messier 20, in Sagittarius

  The Lagoon Nebula and the Trifid Nebula are large, bright H II regions that you can see in the same field of view of your binoculars. Their best observation time is during summer evenings. A color photo shows that the Trifid has a bright red region and a separate, fainter blue region. The red area is the H II region, and the blue zone is a reflection nebula.

Great nebulae of the deep Southern Hemisphere include

- ✔ **The Tarantula Nebula,** in Dorado, the Goldfish

  The Tarantula Nebula is in the Large Magellanic Cloud Galaxy, but it's such a huge and brilliant H II region that it's conspicuous to the naked eye for viewers in temperate southern and far southern latitudes. The Tarantula is another object to observe if you take a South Seas cruise, in addition to the Southern Cross and the Jewel Box star cluster (see the section "Star Clusters: Meeting Galactic Associates," earlier in this chapter).

- ✔ **The Carina Nebula,** in Carina, the Ship's Keel

  The Carina Nebula, near the huge, unstable star Eta Carinae (see Chapter 11), is a large, bright H II region.

- ✔ **The Coal Sack,** in Crux, the Cross

  The Coal Sack, a dark nebula, is a large black patch, several degrees on a side, in the Milky Way. You can't miss it on a clear night with a dark sky, as long as you're deep in the Southern Hemisphere. (By the way, Crux, the Cross, is the official name for what most people call the Southern Cross.)

- ✔ **The Eight-Burst Nebula,** NGC 3132, in Vela, the Sail

  The Eight-Burst Nebula is a planetary nebula that's sometimes called the Southern Ring Nebula; it resembles the Ring Nebula in Lyra, mentioned earlier in this section, but is a bit fainter.

For some of the best color images of nebulae (as well as galaxies and deep space objects) ever photographed, check out the Hubble Heritage Image Gallery, at http://heritage.stsci.edu/gallery/gallery.html.

Another great source for photos of nebulae and other celestial objects is the Astronomy Picture of the Day website, at http://apod.nasa.gov.

Getting a Grip on Galaxies

A large galaxy consists of thousands of star clusters and billions to trillions of individual stars, all held together by gravity. The Milky Way fits this bill as a large spiral system. But galaxies come in many shapes and sizes (see Figure 12-5 for sketches of several main types).

The main types of galaxies, based on shape and size, are as follows:

- ✔ Spiral
- ✔ Barred spiral
- ✔ Lenticular
- ✔ Elliptical
- ✔ Irregular
- ✔ Dwarf
- ✔ Low surface brightness

I cover all these types in the following sections, as well as great galaxies to observe; the Milky Way's home, the Local Group; and even larger groups of galaxies, such as clusters and superclusters.

Figure 12-5: Galaxies come in many different shapes and sizes.

Dinah L. Moché/Astronomy: A Self-Teaching Guide, Seventh Edition

Surveying spiral, barred spiral, and lenticular galaxies

Spiral galaxies are shaped like disks, with spiral arms winding through them. They may resemble the Milky Way, or their arms may be wound more or less tightly than our galaxy's spiral arms. The central bulge of stars in another spiral galaxy may be more prominent or less prominent, compared to the spiral arms. In Figure 12-5, spiral galaxies are labeled by their *Hubble types,* Sa, Sb, and Sc. (Yes, galaxy types are named after Edwin Hubble, too.) As you go from Sa to Sc in that sequence (or beyond, to Sd), the galaxies' spiral arms are less tightly wound and the central bulges are less prominent.

Spiral galaxies have plenty of interstellar gas, nebulae, OB associations, and open clusters, in addition to globular clusters. You can see a photo of spiral galaxies in the color section.

Barred spiral galaxies are spiral galaxies in which the spiral arms don't seem to emerge from the galaxy center; instead, they seem to protrude from the ends of a linear or football-shape cloud of stars that straddles the center. This star cloud is called the *bar*. Gas from outer parts of the galaxy may be funneled toward the center through the bar in a process that forms new stars that make the galaxy's central bulge, well, bulge even more. These galaxies include the types labeled as SBa, SBb, and SBc in Figure 12-5. The sequence from SBa to SBc (and beyond, to SBd, not shown in the figure) goes from barred spirals with tightly wound arms and relatively large central bulges to ones with open arms and small bulges.

Lenticular (lens-shape) galaxies are flattened systems with galactic disks, just like spiral galaxies. They contain gas and dust, but they don't have spiral arms. These galaxies are marked with their type, S0, in Figure 12-5.

Examining elliptical galaxies

Elliptical galaxies are shaped like a football, and that definition includes both the shapes of U.S. footballs and the shapes of soccer balls. Some ellipticals, in other words, are ellipsoidal in shape, roughly like a U.S. football, and some are spherical, like a soccer ball. Ellipticals can be beautiful sights, and I get a kick out of them. They contain many old stars and globular star clusters, but not much else. Elliptical galaxies are shown labeled with Hubble types from E0 to E7 in Figure 12-5. They form a sequence from the roundest ones at E0 to the flattest elliptical galaxies at E7.

A galaxy is a galaxy is a galaxy

Writing *galaxy* and *galaxies* over and over gets repetitive. But what's a good synonym for *galaxy*? Some uninformed folks (or their editors) write *star cluster* to vary their prose, but that's wrong. And a large group of galaxies isn't a *galactic cluster,* an alternate term for an open cluster of stars inside a galaxy. Instead, astronomers call a large group of galaxies a *cluster of galaxies,* sometimes written as *galaxy cluster.* The cluster is composed of galaxies and is thus *galaxian,* but it isn't *galactic.*

Elliptical galaxies are systems in which star formation has largely or totally ceased. They have no H II regions, young star clusters, or OB associations. Imagine living in one of these dull galaxies, with nothing like the Orion Nebula to entertain you or give birth to new stars. And you probably won't find much on television, either.

The production of new stars may have ended in an elliptical galaxy because all the interstellar gas was used up in making the stars already in the galaxy. Or star formation may have ended because something blew out or swept out all the remaining gas suitable for making more stars. I say "suitable" because some elliptical galaxies, although they show no H II regions or groups of young stars, do have some extremely hot gas — so thin and hot, in fact, that it shines only in X-rays. Gas in this state doesn't readily condense into a star. And to tell the truth, some elliptical galaxies do display a number of bluish star clusters, which appear to be very young globular star clusters, much younger than any in the Milky Way.

One leading theory of elliptical galaxies — or at least of *some* elliptical galaxies — is that they form through the collision and merger of smaller galaxies. The collision of two spiral galaxies, for example, can produce a large elliptical galaxy, and shock waves from the event may compress large molecular clouds in the spirals, giving birth to huge clusters of hot, young stars — perhaps the very bluish star clusters found in some ellipticals. But the collision of a small spiral galaxy with a big spiral may lead only to the latter swallowing the former, making the spiral's central bulge even bigger.

As astronomers look out into space, we see many examples of colliding and merging galaxies. The farther out in space we look (so that we are examining the universe at more ancient times), the more prevalent the mergers seem to be. Apparently, galaxy collisions were common in the early universe and may have helped shape many of the galaxies that we see today.

Looking at irregular, dwarf, and low surface brightness galaxies

Irregular galaxies have shapes that tend to be, well, strictly irregular. You may find the glimmerings of a little spiral structure in one of them, or you may not. They generally have plenty of cool interstellar gas, with new stars forming all the time. And they usually appear smaller than full-size spirals and ellipticals, with many fewer stars. You can see an irregular galaxy with its type, Irr, in Figure 12-5.

Dwarf galaxies are just what the name implies: itty bitty galaxies that may be mere thousands of light-years (or less) across. The types of dwarf galaxies include dwarf ellipticals, dwarf spheroidals, dwarf irregulars, and dwarf spirals. Snow White had only seven dwarfs, but the universe may have billions of dwarf galaxies. In our immediate neck of the woods, the Local Group of Galaxies, the most common galaxies are dwarf galaxies — just as in the Milky Way, the most common stars are the smallest stars, red dwarfs. The same probably holds true in the rest of the universe.

Dwarf galaxies are often unusually rich in dark matter, a mysterious substance (or substances) that I describe in Chapter 15.

You don't see dwarf galaxies in Figure 12-5 because Edwin Hubble didn't include them when he made up the original diagram. He didn't include the next type that I list, the low surface brightness galaxies, because they hadn't been discovered yet. Hey, nobody's perfect.

Low surface brightness galaxies were recognized as a major variety in the 1990s. Some of them are as large as most other galaxies, yet they barely shine at all. Although they have a full tank of gas, they haven't produced many stars from the gas, so they don't appear as bright. Astronomers missed them for decades in surveys of the sky, but we're starting to pick them up now with advanced electronic cameras. Astronomers have found some very small low surface brightness galaxies, which are the least luminous galaxies of all. I call them "dim bulb galaxies." Who knows what else is out there that we haven't spotted yet?

Some astrophysicists think that much of the mass in the universe may be present in the form of low surface brightness galaxies that we haven't counted properly, like some groups of people who may be under-represented in the U.S. Census.

Gawking at great galaxies

To enjoy telescopic views of galaxies, use telescopes like those I recommend in the section "Enjoying Earth's best nebular views," earlier in this chapter. Big galaxies like the Andromeda Galaxy or the Triangulum Galaxy are great sights through a low f/number telescope (see Chapter 3). For viewing smaller galaxies, I recommend a telescope with a computer control feature that points the telescope to exactly the right place in the sky. *Sky & Telescope's Pocket Sky Atlas* and other atlases show where the bright galaxies are among the constellations.

The best galaxies for viewing from the Northern Hemisphere include the following. When I mention the season of the year for best viewing, I mean the season in the Northern Hemisphere. (Remember, it's fall in the Northern Hemisphere when Brazilians are enjoying spring.)

- **The Andromeda Galaxy** (Messier 31; see Chapter 1), in Andromeda, a constellation named for an Ethiopian princess in Greek mythology

  The Andromeda Galaxy is also called the Great Spiral Galaxy in Andromeda, and it was long known as the Great Spiral Nebula in Andromeda, or just the Andromeda Nebula. It looks like a fuzzy patch to the naked eye. You can see it in the autumn evening sky. From a dark observing site, you can trace it across about 3° — or about six times the width of the full Moon — on the sky with your binoculars. Don't try to view this galaxy during the full Moon — you will get a poor view, at best; wait until the Moon is just a crescent or is below the horizon. The darker the night, the more of the Andromeda Galaxy you can see.

- **NGC 205 and Messier 32,** in Andromeda

  NGC 205 and Messier 32 are small, elliptical companion galaxies of the Andromeda Galaxy. Some experts call them both dwarf elliptical galaxies, and some don't. (I wish they would make up their minds.) M32 is spheroidal in shape, and NGC 205 is ellipsoidal.

- **The Triangulum, or Pinwheel Galaxy** (Messier 33), in Triangulum, the Triangle

  The Triangulum, or Pinwheel Galaxy, is another large, bright, nearby spiral galaxy, smaller and a little dimmer than the Andromeda Galaxy — and also a fine sight through binoculars in the fall.

- **The Whirlpool Galaxy** (Messier 51), in Canes Venatici, the Venetian Hunting Dogs (see Figure 12-6)

  The Whirlpool Galaxy is farther away and fainter than the Andromeda Galaxy and the Triangulum Galaxy, but you get a more glorious view of it through a high-quality small telescope. The Whirlpool is a *face-on* spiral, meaning that the galactic disk is pretty much at right angles to our line of sight from Earth; we look right down (or up) on it. With the larger telescopes at a star party (see Chapter 2), you should be able to make out its spiral structure from a distance of about 23 million light-years. Messier 51 is where the 3rd Earl of Rosse discovered the spiral structure of galaxies in 1845. (He just happened to own the largest telescope in the world.) Look for it on a fine, dark evening in the spring.

✔ **The Sombrero Galaxy** (Messier 94), in Virgo, the Virgin

The Sombrero Galaxy looks like a bright, edge-on spiral galaxy, and that's how astronomers classified it until recently. However, the latest theory is that the Sombrero is a giant elliptical galaxy that somehow has a structure resembling a spiral galaxy inside it. The "brim" of the Sombrero is the spiral structure's galactic disk. A dark stripe appears along the brim because the band of dark nebulae or coal sacks in the disk is edge-on to our line of sight. Look for the Sombrero in the spring, too; it's a bit farther away than the Whirlpool, but it still looks good through a telescope.

The following list presents the finest galaxies for observers in the Southern Hemisphere:

✔ **The Large and Small Magellanic Clouds** (LMC and SMC) are irregular galaxies that orbit the Milky Way. The Large Cloud is not only larger, but also closer to Earth. It orbits a mere 169,000 light-years (give or take) from us. In fact, scientists believed for many years that the LMC was the closest galaxy to the Milky Way. (Today scientists know that two dim, miserable excuses for a galaxy, called the Sagittarius Dwarf Galaxy and the Canis Major Dwarf Galaxy, are even closer. However, we can barely discern those two objects in telescopic photos because the Milky Way is absorbing them.)

The LMC and SMC actually look like clouds in the night sky. They're that big and bright and are circumpolar in much of the Southern Hemisphere. In other words, at far southern latitudes, they never set below the horizon. If you go far enough south in South America or elsewhere in the Southern Hemisphere, the LMC and SMC are visible on every clear night of the year. Sweep through them with binoculars and see how many star clusters and nebulae you can recognize.

Figure 12-6:
The Whirlpool Galaxy, photographed in ultraviolet light by the GALEX satellite.

Courtesy of NASA/JPL/Caltech

✔ **The Sculptor Galaxy** (NGC 253) is a large, bright spiral galaxy and is one of the dustiest. Caroline Herschel, who also found eight comets, discovered it in 1783. Southern Hemisphere folks can look for it with binoculars or a telescope on a dark spring evening. Observers in the continental United States who have an unobstructed southern horizon can look for it low in the sky in autumn.

✔ **Centaurus A** (NGC 5128) is a huge galaxy with a peculiar appearance: spheroidal, but with a thick band of dark dust across its middle. The galaxy is a powerful source of radio waves and X-rays and has been much studied with radio telescopes and with X-ray telescopes in orbiting satellites. Theorists have gone back and forth on whether it's an example of colliding galaxies. Some astronomers suspect that it may be another object like the Sombrero Galaxy: a giant elliptical galaxy with spiral structure within. Regardless of which theory is correct, I think that Centaurus A has swallowed a smaller galaxy or two in its time, so watch from a safe distance. This object is most suitable for viewing during autumn in the Southern Hemisphere.

Discovering the Local Group of Galaxies

The Local Group of Galaxies, called the Local Group for short, has more than 50 members. It includes two large spirals (the Milky Way and the Andromeda Galaxy), a smaller spiral (the Triangulum Galaxy), their satellites (including the Large and Small Magellanic Clouds, as well as M32 and NGC 205), and lots of dwarf galaxies.

The Local Group isn't much as assemblages of galaxies go, but it's home and the largest structure that we on Earth are gravitationally bound to (meaning that Earth isn't flying away from the Local Group as the universe expands). Just as the solar system isn't getting any bigger — because the Sun's gravity prevents the planets from moving outward or escaping — the Local Group holds strong because of the gravity of the three spiral galaxies and the smaller members. But all other groups and clusters of galaxies and distant individual galaxies *are* moving away from the Local Group at rates determined by a formula called *Hubble's Law* (named for You-Know-Who). Chapter 16 explains more about the away-ward movement.

The Local Group is about 1 megaparsec across and centered near the Milky Way. A *parsec* is a dimension in space equal to 3.26 light-years, and *mega* means "million," so the Local Group is about 3.26 million light-years, or about 19 trillion miles, wide. That dimension may sound large, but some experts think it's even bigger. Either way, the Local Group is minuscule compared to the vast extent of the observable universe.

Clusters and superclusters of galaxies are much larger than the Local Group, easily spotted across billions of light-years in space. But the majority of all the galaxies in the universe, at least the readily visible ones, are located in small groups with only dozens of members or less, like the Local Group (which has about 30). So we appear to be in an average condition, as galaxy neighborhoods go.

Checking out clusters of galaxies

Most galaxies may be in small groups like the Local Group, but as astronomers survey the distant heavens with professional observatory telescopes, the formations that stand out are the clusters of galaxies. Most prominent are the so-called *rich clusters,* with hundreds and even thousands of galaxy members, each with its own complement of billions of stars.

The nearest large cluster of galaxies is the Virgo Cluster, spread out across the constellation of the same name and adjacent constellations. The cluster is about 54 million light-years away and contains more than 1,000 galaxies.

You can observe some of the biggest and brightest member galaxies of the Virgo Cluster with your own telescope. Messier 87 is one of the best sights: a spheroidal giant elliptical galaxy with a powerful jet of matter flying out at its center from the vicinity of a supermassive black hole. You can see M87 with amateur equipment, but not the jet at its center, unless you're a *very* advanced amateur. The galaxy appears to have swallowed some smaller ones, which may be why it's so big. Some galaxies like to start small and work their way up. Messier 49 and Messier 84 are two more Virgo Cluster giant ellipticals that you can observe, and Messier 100 is a large spiral galaxy in the cluster. Look for these galaxies on a dark evening in the spring in the Northern Hemisphere. Use a telescope with computer control that can home in on them for you. And if you don't trust computers, be sure you have a good star atlas that shows the galaxies.

Clusters of galaxies exist as far as our telescopes can see. At the limit of current technology in the early 21st century, we estimate that there are about 100 billion galaxies in the observable universe, but nobody has counted them all — at least, nobody on our planet.

Sizing up superclusters, cosmic voids, and Great Walls

You may think that a large cluster of galaxies, up to 3 million light-years across, would be the max. But deep-sky surveys indicate that most or all galaxy clusters are grouped into larger forms, called *superclusters.* The superclusters aren't held together by gravity, but they haven't fallen apart, either.

They appear to have long, filamentary shapes and flat, pancakelike shapes. A supercluster can contain a dozen clusters of galaxies, or hundreds of these clusters, and it can be 100 or 200 million light-years long.

We exist in the outer parts of the Local Supercluster, sometimes called the Virgo Supercluster, which is centered near the Virgo Cluster of Galaxies.

The superclusters seem to be positioned on the edges of huge, relatively empty regions of the universe called *cosmic voids*. The nearest one, the Bootes Void, is more than 300 million light-years across. Many galaxies sit on its periphery, but we don't see very many inside the Void.

Astronomer Robert Kirshner discovered the Bootes Void. But when he was congratulated on the find, he reportedly said modestly, "It's nothing."

Some of the biggest superclusters, or groups of superclusters, are called *Great Walls*. The first Great Wall discovered is about 750 million light-years long. But other Great Walls, far out in the universe, may be larger. As far as astronomers know, the Great Walls don't display any Great Graffiti, but they have plenty to tell us about the origin of large structures in space and the early history of the universe. If only we understood the language.

Joining Galaxy Zoo for Fun and Science

Now that you know the basic types of galaxies, why not help astronomers sift through the numerous beautiful pictures of galaxies from the Hubble Space Telescope? As in other citizen science projects that I describe in the final section of Chapter 11, all you need is your brain and your computer, with access to the Internet.

By signing up for the Galaxy Zoo at `www.galaxyzoo.org`, you add your efforts to those of more than a quarter-million people who have helped professional astronomers study distant galaxies captured in images from a ground-based sky survey. That work is finished, and now the Zookeepers are asking for help with images from Hubble.

After enrolling in Galaxy Zoo, you study examples of galaxy images and learn how to classify them. Then you're set to scan the Hubble pictures and help astronomers uncover new facts about the universe. In Galaxy Zoo's first year, almost 150,000 volunteers, called *Zooites*, provided more than 50 million galaxy classifications, judging whether galaxies were spirals or ellipticals. More recently, reports from Zooites overturned the earlier belief that nearly all elliptical galaxies are red. Their findings showed that about one-third of all red galaxies are spirals.

One Galaxy Zoo volunteer, Dutch teacher Hanny Van Arkel, discovered an object in Leo Minor (the constellation of the Little Lion) that didn't look like any known type of galaxy. She has gone down in the history of astronomy as the discoverer of this strange object, called *Hanny's Voorwerp* (Dutch for "Hanny's Object"). Ms. Van Arkel has been lionized by astronomers. Now that's something to roar about!

As an added benefit, you get to peer at the Hubble pictures for free! You have to pay to visit most major planetariums, but there's never a charge at the Galaxy Zoo.

Chapter 13

Digging into Black Holes and Quasars

*B*lack holes and quasars are two of the most exciting and sometimes mystifying areas of modern astronomy. Lucky for us astronomers, the two subjects are related. In this chapter, I explain the connection between the two mysteries and provide you with information about active galactic nuclei, a group that quasars fall into.

You may never see a black hole through your own telescope, but I can guarantee that when you tell people you're an astronomer, they'll blurt out, "What's a black hole?" I mention black holes briefly in Chapter 11, but in this chapter, I offer you the whole treatment.

Black Holes: Keeping Your Distance

A *black hole* is an object in space whose gravity is so powerful that not even light itself can escape from within — which is why black holes are invisible.

You can fall into a black hole, but you can't fall out — you can't get out if you want to (and you *would* want to). You can't even call home, so E.T. (in the 1982 movie) was lucky he landed in California and not in a black hole.

Anything that enters a black hole needs more oomph than it can ever have to get back out. The formal name for that "oomph" is escape velocity. Rocket scientists use the term *escape velocity* to represent the speed at which a rocket or any other object must travel to escape Earth's gravity and pass into interplanetary space. Astronomers apply the term in a similar way to any object in the universe.

The escape velocity on Earth is 7 miles per second (11 kilometers per second). Objects with weaker gravity have slower escape velocities (the escape velocity on Mars is only 3 miles or 5 kilometers per second), and objects with more powerful gravity have higher escape velocities. On Jupiter, the escape velocity is 38 miles per second (61 kilometers per second). But the champion of the universe for escape velocity is a black hole. The gravity of a black hole is so strong that its escape velocity is greater than the speed of light (186,000 miles per second or 300,000 kilometers per second). Nothing, not even light, can escape from a black hole (because you need to travel faster than the speed of light to escape a black hole, and nothing — including light — travels faster).

In 2011, a team of physicists reported on an experiment in which they found indications that some neutrinos (a type of subatomic particle that I describe in Chapter 10) travel faster than the speed of light. Such a finding would contradict a certain amount of known physics if it were correct, but it wasn't. Experts later traced the error in the experiment to a loose electrical connector. The physicists didn't have a screw loose, but the situation was almost as bad.

Some scientists have proposed a hypothetical class of particles, called *tachyons,* that can move faster than light. In fact, if tachyons exist, they can never travel at *less* than light speed. But the theory that these particles exist isn't widely accepted, and none of them have been found.

Looking over the black hole roster

Scientists can detect black holes when they see gas swirling around them that's too hot for normal conditions, when jets of high-energy particles make their escape as though to avoid falling in (the jets come not from inside a black hole, but from the black hole's immediate surroundings), and when stars race around orbits at fantastic speeds, as though driven by the gravitational pull of an enormous unseen mass (which they are).

As I mention in Chapter 11, scientists recognize two main types of black holes:

- **Stellar mass black holes:** These black holes have the mass of a large star (about three to a hundred times more massive than the Sun), and they result from the deaths of such stars.

- **Supermassive black holes:** These whoppers range from hundreds of thousands to more than 20 *billion* times as massive as the Sun and exist at the centers of galaxies. They may have come from the mergers of many closely packed stars or the collapse of huge gas clouds when the galaxies formed. But nobody knows for sure.

In 1999, scientists also discovered intermediate mass black holes, which have masses of a few hundred to perhaps 10,000 times that of the Sun, but they don't know how these black holes formed.

Poking around the black hole interior

A black hole has three parts:

- **The event horizon:** The perimeter of the black hole
- **The singularity:** The heart of the hole formed from the ultimate compression of all matter within it
- **Falling objects:** Matter that falls from the event horizon toward the singularity

The following sections describe these parts in more detail.

The event horizon

The event horizon is a spherical surface that defines the black hole (see Figure 13-1). After an object enters the event horizon, it can never get back out of the black hole or be visible again to anyone on the outside. And nothing on the outside can be seen from inside the horizon.

The size of the event horizon is proportional to the mass of the black hole. Make the black hole twice as massive, and you make its event horizon twice as wide. If scientists had a way to squeeze Earth until it became a black hole (we don't, and if we did, I wouldn't tell how), our planet would have an event horizon less than ¾ of an inch (about 2 centimeters) across.

Figure 13-1: One concept of a black hole, with arrows representing doomed matter falling in.

Table 13-1 offers a list of black hole sizes, in case you want to try some on. The two largest black holes in the table are at the centers of giant elliptical galaxies that are the brightest and most massive galaxies in the respective clusters of galaxies in which they're located. (I describe clusters of galaxies in Chapter 12.)

Table 13-1	Black Hole Measurements		
Black Hole Mass in Solar Masses	*Diameter of Black Hole in Miles*	*Diameter of Black Hole in Kilometers*	*Comment*
3	11	18	Smallest stellar mass black hole
10	37	60	Typical stellar mass black hole
100	370	600	Largest stellar mass black hole
1,000	3,700	6,000	Intermediate mass black hole
4 million	15 million	24 million	Supermassive black hole at Milky Way center
6.3 billion	23 billion	37 billion	Supermassive black hole in M87 in the Virgo Cluster
21 billion	78 billion	130 billion	Supermassive black hole in NGC 4889 in the Coma Cluster

As far as scientists know, no black holes are smaller than about three solar masses and 11 miles wide.

The singularity and falling objects

Anything that falls inside the event horizon moves down toward the singularity. It merges into the singularity, which scientists believe is infinitely dense. We don't know what laws of physics apply at these immense densities, so we can't describe what the conditions are like. We literally have a black hole in our knowledge.

Some mathematicians think that at the singularity there may be a *wormhole,* a passage from the black hole to another universe. The wormhole concept has inspired authors and movie directors to produce plenty of science fiction on the topic, but the writers and directors are just fishing. Most experts think that wormholes don't exist. And even if they do, scientists have no way to see wormholes inside black holes or to inch our way down to them.

Another theory says that, at the location where the hypothetical wormhole connects to another universe, there's a *white hole* — a place where enormous energy pours out into the other universe, like a gift from our universe. This idea, too, seems wrong, but even if the theory *is* correct, we would have to journey to another universe (talk about frequent-flyer miles!) to see one.

Traveling to another universe (if one exists) is out of the question. But the other possibility, of course, is to look for white holes in our universe, where a wormhole from another universe may emerge. Scientists have found no such thing.

Someone once suggested that bright quasars may be white holes. But today astronomers have perfectly good explanations for quasars (as I explain in the section "Quasars: Defying Definitions," later in this chapter), so as far as I'm concerned, astronomers are off the hook.

Surveying a black hole's surroundings

Here's what scientists observe in the vicinities of black holes:

1. **Gaseous matter falling toward the black hole swirls around in a flattened cloud called an *accretion disk*.**

2. **As the gas in the accretion disk gets closer to the black hole, it becomes denser and hotter.**

   The gas heats up because the gravity of the black hole compresses it, a process that occurs because friction increases as the gas gets denser. (The process resembles the way air conditioners and refrigerators work: When gas expands, it cools, and when it gets compressed, it gets hotter.)

3. **As the hotter and denser gas approaches the black hole, it glows brightly; in other words, the accretion disk shines.**

   Radiation from the accretion disk can take many forms, but the most common type is X-rays. X-ray telescopes, such as the one onboard NASA's large Earth-orbiting satellite, the Chandra X-ray Observatory, detect X-rays and allow scientists to pinpoint black holes. You can see the X-ray images from Chandra at `http://chandra.harvard.edu`, the website of the Chandra X-ray Center, by clicking on the Photo Album link.

So although you can't actually see a black hole through a telescope, you can detect radiation from the accretion disk of hot gas swirling around it — if you have an X-ray telescope that's up in space. X-rays don't penetrate Earth's atmosphere, so the telescope needs to be above the atmosphere.

Bare black holes may exist in space, with no gas swirling into them. If so, astronomers can't see them unless they just happen to pass in front of a background star or galaxy under observation. In that case, you can infer that the black hole exists because you see the effect of its gravity on the appearance of the background object. (You may see the background object get briefly brighter, for example, as I describe in Chapter 11 when I discuss gravitational microlensing.) But this situation is a rare coincidence. Don't hold your breath waiting for a black hole to bare itself for your inspection.

Warping space and time

You can think of a black hole as a place where the fabric of space and time is warped. A *straight line* — which is defined in physics as the path taken by light moving through a vacuum — becomes curved in the vicinity of a black hole. And as an object approaches a black hole, time itself behaves oddly, at least as perceived by an observer at a safe distance.

Suppose that, as you move at a safe distance, you launch a probe toward a black hole from your spaceship. A big electric billboard on the side of the probe displays the time given by a clock in the probe.

From the spaceship, you watch the clock through a telescope as the probe falls toward the black hole. You see that the clock runs slower and slower as the probe approaches the black hole. In fact, you never actually see the probe fall in. You see it get redder and redder as the light from the electric billboard is redshifted by the powerful gravity of the black hole — not because of the Doppler effect (which I describe in Chapter 11), but because of a phenomenon called *gravitational redshift.* The light of the billboard shifts toward longer wavelengths, just as the Doppler effect makes the light from a star moving away from the observer seem to shift toward longer wavelengths. After awhile, gravity redshifts the glow of the electric billboard to infrared light, which your eyes can't detect.

Now consider what you'd see if you traveled onboard the falling probe. (Don't try this at home — in fact, don't try it anywhere.) You can watch the clock inside the probe, and you can peek back along your path through a window. You, the ill-fated onboard observer, see that the clock runs normally. You don't perceive that it runs slow at all. As you look out the window at the mother ship and the stars, they all seem to be blueshifted. You're blue at the thought that you can never go home again. You pass through an invisible boundary (the event horizon) around the black hole in no time. From then on, you can never see what's outside the event horizon again, nor can anyone outside see you.

A person on the mother ship never sees you enter the black hole; you just appear to get closer and closer to the hole. But on the falling probe, you can tell that you've dropped right in. At least, you can if you're still alive. Ultimately, *tidal force,* an effect of the immense gravity, pulls apart anything that falls into a black hole; at least, along one dimension (the direction toward the singularity), you're pulled apart. To make matters worse, in the other two spatial dimensions, tidal force squeezes you together unmercifully.

If you enter the black hole feet first, tidal force stretches you out (if you're not already pulled apart) until you are more than tall enough to be drafted as a center in the National Basketball Association. But from bellybutton to back and from hip to hip, you get squeezed together like coal turning into diamond under immense pressure inside Earth, only worse. This experience is no gem.

Small or stellar mass black holes are the most deadly variety, just as some small spiders are more poisonous than big tarantulas. If you fall toward a stellar mass black hole, you're torn apart and squeezed together before you enter, and you never get to see the universe disappear before you die. But falling into a supermassive black hole is a different experience. You get to fall inside the event horizon and see the universe black out before you suffer the tidal fate (or is it the fatal tide?).

Considering that black holes are all around us in the universe and that they have such fascinating and strange properties, you can see why scientists want to study them, but from a safe distance.

Watching stars get swallowed by black holes

The tale of a person who may be ripped apart by a black hole (as I describe in the preceding section) is hypothetical. But the tidal force of a black hole is a real threat that can tear almost anything to pieces. The exception is that a black hole doesn't tear apart another black hole. Instead, the two black holes merge, making a single, bigger black hole. If a star comes too close to a really large supermassive black hole, one with a mass of about 100 million times the mass of the Sun or more, the black hole swallows it before it tears apart the star. So sometimes a star ripped up by a black hole can be observed from afar, and sometimes it cannot.

Astronomers estimate that, in a typical galaxy with a supermassive black hole in its center, a star comes near enough to be ripped open by the black hole in a manner that we can observe about once every 100 centuries. You may need to stare at a galaxy that long before you witness a *tidal disruption event,* the technical term for a star's death by black hole. Because no one has that much time on their hands, observers use robotic telescopes equipped with large digital cameras to monitor thousands of galaxies. They're looking for bright flares — increases in brightness that may last weeks or months — that signal the tidal breakups of stars. If you keep tabs on 10,000 galaxies with the robotic equipment, you can expect to catch about one flare per year.

Astronomers have detected several likely tidal disruption events in recent years. A good example occurred on May 31, 2010, when a prototype telescope for the Panoramic Survey Telescope & Rapid Response System (Pan-STARRS; http://pan-starrs.ifa.hawaii.edu/public) on Mount Haleakala, in Hawaii, found a flare underway in a dim, nameless galaxy about 3 billion light-years from Earth in the constellation Draco. The Pan-STARRS observers made precise measurements of the flare as the brightness rose and fell, until it was last seen on September 1 of the following year. They also obtained observations made in ultraviolet light by NASA's Galaxy Evolution Explorer satellite (check out its website at www.galex.caltech.edu). Putting together all this information, the researchers drew the following conclusions:

- A central black hole of about 3 million times the mass of the Sun ripped apart a star that had begun life as a star much like the Sun.

- The ill-fated star became a red giant (as the Sun will do some day — see Chapter 10) and then lost its outer layers. The missing layers, composed mostly of hydrogen, may have been pulled off earlier by the black hole. What was left of the star before the flare began was its dense core of helium, with a mass of roughly one-quarter of the mass of the Sun.

- During the tidal disruption event, half of the star's core was lost to the black hole, and the other half was flung away into the surrounding galaxy.

The black hole at the center of the Milky Way galaxy has a mass of 4 million Suns, more than the black hole that caused the star disruption observed by the Pan-STARRS prototype telescope. So what happened in the nameless galaxy 3 billion light-years from Earth can happen right in our own corner of the universe. In fact, a tidal disruption event in the Milky Way is underway as this book goes to press. But the victim, in this case, is a nebula. Astronomers observing with an 8-meter telescope at the European Southern Observatory (www.eso.org/public) in Chile have been watching a small nebula fall in toward the Milky Way's central black hole, Sagittarius A*, since 2008. The nebula is about three times as wide as Pluto's average distance from the Sun, and its mass equals about three times the mass of Earth. The tidal force of the black hole is stretching the nebula along the direction in which it's falling and is squeezing it in at right angles to that direction.

The astronomers who found the doomed nebula estimate that its orbit will take it closest to the black hole in mid-2013, when it will likely be disrupted, with some nebular matter falling into the black hole's accretion disk and producing what can be a bright flare.

Quasars: Defying Definitions

Scientists have at least two definitions for quasars:

- ✔ **The original definition:** *Quasar* is an abbreviation or acronym for "quasi-stellar radio source" and means a celestial object that emits strong radio waves but looks like a star through an ordinary visible light telescope (see Figure 13-2).

  The original definition of *quasar* has become outdated because, at most, 10 percent of all objects that we now call quasars fit this definition. The other 90 percent don't emit strong radio waves. Astronomers call them radio-quiet quasars.

- ✔ **The current definition:** A *quasar* is an *active galactic nucleus,* meaning a supermassive black hole with an accretion disk that is fed by inflowing matter from the surrounding galaxy.

The current definition of *quasar* reflects the fact that, after decades of puzzling over quasars, astronomers have concluded that they're associated with giant black holes at the centers of galaxies. Matter that falls into the black hole releases enormous energy, and the observed energy sources are what astronomers call quasars.

Figure 13-2:
A quasar shines with 10 trillion times the power of the Sun.

Courtesy of NASA

The flow of matter into the black hole in a quasar can vary. When the flow is ample, the quasar shines with as much as 10 trillion times the energy the Sun releases each second. When the flow is low, however, the quasar fades away until it is revived with a good meal at a later time.

Measuring the size of a quasar

All quasars produce strong X-rays; about 10 percent produce strong radio waves; and they all emit ultraviolet, visible, and infrared light. All the emissions can vary in strength over years, months, and weeks, and even over times as short as a day.

The fact that quasars often change significantly in brightness over the course of a single day indicates to scientists something of surpassing importance: A quasar must be no larger than about one *light-day,* or the distance that light travels through a vacuum over the length of a day. And a light-day is only 16 billion miles (26 billion kilometers) long, which means that a quasar, producing as much light as 10 trillion suns, or 100 times as much light as the Milky Way Galaxy, isn't much bigger than our solar system, which is a tiny part of the Galaxy.

A quasar much bigger than a light-day can't fluctuate markedly in such a short time, any more than an elephant can flap its ears as rapidly as a hummingbird flutters its wings.

Getting up to speed on jets

Quasars that are strong radio sources often display *jets,* or long, narrow beams in which energy shoots out of the quasars in the form of high-speed electrons and perhaps other rapidly moving matter. Often the jets are lumpy, with blobs of matter moving outward along the beams. Sometimes the blobs appear to move at more than the speed of light. This *superluminal motion* is an illusion related to the fact that the jets, in such cases, are pointing almost exactly at Earth; the matter in them actually moves close to light speed, but not faster than light.

You can browse the best images of jets from quasars as detected by radio telescopes in the Image Gallery of the National Radio Astronomy Observatory, at `http://images.nrao.edu`.

Exploring quasar spectra

Many books say that a quasar has very broad lines in its spectrum, corresponding to redshifts and blueshifts of gas in turbulent motion within the quasar at up to 6,000 miles per second (10,000 kilometers per second). This statement isn't always true. Quasars come in a variety of types, and some don't have broad spectral lines. (See Chapter 11 for more on spectral lines.)

The broad spectral lines, however, are an important trait of many quasars and are a clue to their relationship to other objects, as I describe in the next section.

Active Galactic Nuclei: Welcome to the Quasar Family

For years after the discovery of quasars, astronomers argued about whether they're located in galaxies or are separate from galaxies. Today we know that quasars are indeed always located in galaxies because technology has improved to the point that we can make a telescopic image that shows both a quasar and the galaxy around it. The latter is called the quasar's *host galaxy*. (Previously, because a quasar can be 100 times brighter than its host galaxy — or even brighter — hosts got lost in the glare of their quasar guests in telescopic photos.)

Electronic cameras, which can record a greater range of brightnesses in a single exposure than photographic film, made the discovery possible.

Quasars are an extreme form of what astronomers now call *active galactic nuclei (AGN)*. The term designates the central object of a galaxy when the object has quasarlike properties, such as a very bright starlike appearance, very broad spectral lines, and detectable brightness changes.

Sifting through different types of AGN

Scientists use the following main terms to describe AGN:

✔ **Radio-loud quasars ("original quasars") and radio-quiet quasars (90 percent or more of quasars):** These two types are similar kinds of objects, with and without strong radio emission. They're located in spiral galaxies (see Chapter 12 for more on galaxies). No quasar is visible in the Milky Way Galaxy, but we have detected a 4-million-solar-mass black hole at its center, called Sagittarius A*. I list that supermassive black hole in Table 13-1.

✔ **Quasistellar objects (QSOs):** Some astronomers lump together the radio-loud and radio-quiet quasars as QSOs.

✔ **OVVs:** *Optically violently variable quasars* are quasars with jets that happen to point directly at Earth. These quasars undergo even more pronounced rapid brightness changes than ordinary, garden-variety quasars. Think of firefighters struggling to direct a firehose at a person whose clothes are on fire. The water pressure may be unstable, with the water pulsing a bit. The stream from the hose may look pretty steady to spectators from the side, but the person on the receiving end feels every fluctuation in the flow as the oncoming water batters him. OVVs are the firehoses of quasardom — they make the biggest splash.

✔ **BL Lacs:** Astronomy lingo for *BL Lacertae objects,* BL Lacs, as a group, are AGN that resemble BL Lacertae. BL Lacertae changes in brightness; for years, scientists thought it was just another variable star in the constellation Lacerta (it looks like a star in photographs of the sky). They later identified it as a strong source of radio waves and eventually determined that BL Lacertae was the active nucleus of a host galaxy that had been lost in its glare until improved technology made it possible to photograph the galaxy.

Unlike most quasars, a BL Lac doesn't have broad spectral lines. And its radio waves are more highly polarized than those from ordinary radio-loud quasars (except the OVVs, which may be just extreme cases of BL Lacs) — *polarized* means that the waves have a tendency to vibrate in a preferred direction as they travel through space. Unpolarized waves vibrate equally in all directions as they move. At the ballpark, you can't tell the players without a scorecard, as the saying goes; at the observatory, you have to check polarization to know your radio-loud quasars from your BL Lacs.

✔ **Blazars:** This term covers both OVVs and BL Lacs. The two quasar types have many similarities. Both are highly variable in brightness, their jets point directly at Earth, and they're both radio-loud.

Do we really need a term to combine OVVs and BL Lacs? I'm not so sure. My friend Dr. Hong-Yee Chiu became famous among scientists for coining *quasar.* His friend Professor Edward Spiegel coined *blazar* a few years later. If you discover a new kind of object or write one of the leading studies on it, you may get to name it, too. Adding *ar* to your name isn't allowed; the term should be descriptive of the scientific properties of the object, not the astronomer.

✔ **Radio galaxies:** These galaxies have relatively dim active galactic nuclei that nevertheless produce strong radio emissions. Most of the strongest radio-emitting galaxies are giant elliptical galaxies. Often they have beams or jets that transport energy from the AGN to huge lobes of radio emission, empty of stars, far outside and far larger than the host galaxy itself. Usually there is a lobe on one side of the galaxy and a second lobe on the opposite side.

✔ **Seyfert galaxies:** These spiral galaxies have an AGN at their centers. A Seyfert AGN is like a quasar, with broad spectral lines and rapid brightness changes. It may be as bright as the host galaxy, but not 100 times brighter like a quasar, so the host isn't lost in the Seyfert nucleus's glare.

A Seyfert nucleus isn't a demanding guest; it's more like a minor presidential candidate who visits a small town without causing a big fuss. Local folks know that the candidate is around, but they tend to follow their daily routines instead of flocking downtown to greet the visitor. Carl Seyfert was the American astronomer who pioneered the study of these galaxies and their bright centers, long before astronomers knew that black holes exist.

Examining the power behind AGN

All the different types of AGN have one thing in common: They're powered by energy that's generated in the vicinity of the supermassive black holes at their centers.

Near the supermassive black hole, stars orbit the center of the host galaxy at immense speeds, a circumstance that enables astronomers to measure the black hole's mass. With telescopes such as the Hubble, astronomers can determine the velocities of the orbiting stars or sometimes orbiting gas clouds by measuring Doppler shifts of the light from the stars or the gas (see Chapter 11 for more about the Doppler effect). The speeds indicate the mass of the central object. Stars at the same distance from the center of a less-massive black hole orbit at a slower pace.

In a quasar or a radio galaxy of the giant elliptical type, the supermassive black hole often attains a billion or more solar masses. In Seyfert galaxies, the black hole mass is often around a million solar masses.

The black hole makes it possible for the AGN to shine, but only the matter falling into the black hole actually powers the glow. It may take matter with ten times the mass of the Sun falling into the black hole each year to make a quasar shine.

If no material falls into the black hole, the AGN doesn't reveal itself by producing a bright glow, radio emission, high-speed jets, or strong X-rays. Similar to kids who depend on their school lunches for the energy to perform in class, the black holes shine only when matter falls into them at a sufficient rate. Supermassive black holes may be lurking at the centers of most galaxies, but in most cases, matter isn't feeding them, so astronomers see quasars or other kinds of AGN in only a small fraction of galaxies.

Proposing the Unified Model of AGN

The *Unified Model of Active Galactic Nuclei* is a theory proposing that many kinds of AGN are, in fact, the same kind of object, which looks different when viewed from different angles. According to the Unified Model, when we look at AGN from different directions with respect to their accretion disks and their jets, they look different, just as a man you see face-on looks different than the same guy in profile. Everyone has a good side; from some angles, Jay Leno's chin doesn't seem so big. The theory also proposes that black holes are sucking in matter at different rates, so some AGN (which are getting more matter per second than others) are brighter than the others for that reason alone. Dozens of astronomers write papers on the Unified Model every year; some find evidence for the theory, and some find evidence against it.

I think that evidence points to real differences among the different types of AGN, but I also think they have many basic similarities. Astronomers need more information before we can unite around the Unified Model or any other theory of AGN. In the meantime, what do you think? Your taxes pay for much of this research, which is underway in just about every developed nation, so you're entitled to an opinion.

What came first: The black hole or the galaxy?

An important discovery brightened the day of every quasar fan. Experts found a simple mathematical relationship between a supermassive black hole and the galaxy that surrounds it. The central region of most galaxies is called the *bulge.* Even a relatively flat spiral galaxy may have a central bulge, which can be big, middling, or small. An elliptical galaxy is considered all bulge. Astronomers have found that the mass of a black hole at the center of a bulge is always close to one-fifth of 1 percent of the mass of the whole bulge. It seems as though every galaxy has to pay a tax of 0.2 percent to its black hole. (I wish the Internal Revenue Service let me off that cheaply.)

This unexpected property of black holes and galaxies must be related to how they form, but astronomers aren't sure how. Does a big galaxy form around a big black hole? Or do big black holes form inside big, bulgy galaxies? Astronomers are arguing over this now in a dispute that I call the "battle of the bulges."

Part IV
Pondering the Remarkable Universe

The 5th Wave — By Rich Tennant

Kenny didn't need the distraction, but it was just his luck that the Zorlocks' invasion of Earth began JUST as he started his Astronomy test.

In this part . . .

Read Part IV when you need a diversion, something to stir your mind with thought-provoking ideas and possibilities. Curl up with a glass of wine or a cup of cider and read about the Search for Extraterrestrial Intelligence (SETI). Have scientists discovered any evidence that little green beings are out there? Find out about dark matter, dark energy, and antimatter (yes, antimatter exists in the real world, not just in science fiction). And when you feel ready, ponder the entire universe: how it began, its shape, and its future.

Chapter 14

Is Anybody Out There? SETI and Planets of Other Suns

• •

In This Chapter

▶ Understanding the Drake Equation

▶ Exploring (and participating in) SETI projects

▶ Hunting for extrasolar planets and seeing what they're like

• •

*T*he universe is both vast and varied. But do we share these starry realms with other thinking beings? Anyone who tunes in to *Star Trek* or frequents the local Cineplex already knows Hollywood's answer: The cosmos is cluttered with aliens (many of whom have managed to pick up a good amount of uninflected English).

But what do scientists say? Are there really extraterrestrial beings out there? Most researchers believe that the answer is yes. Some are even looking for evidence. Their quest is known as SETI (rhymes with "Betty"), the Search for Extraterrestrial Intelligence. (Other scientists have plans to search for traces of primitive life on Mars or some of the moons of the outer solar system, but SETI seeks advanced civilizations capable of broadcasting into space.)

Why are many scientists optimistic about the possible existence of aliens? Most of the upbeat attitude derives from the fact that our place in the cosmos is thoroughly unremarkable. The Sun may be an important star to us, but it serves as a bit player in the universe. The Milky Way galaxy hosts 10 billion similar stars. If this number fails to impress you, note that more than a hundred billion *other* galaxies are within range of our telescopes. The bottom line is that far more Sun-like stars are sprinkled through the visible universe than we have blades of grass on Earth. To assume that our blade of grass is the only place where something interesting happens is (to put it gently) a bit gutsy. Distressing as it may be to our self-esteem, Earth may not be the intellectual nexus of the universe.

How can earthlings find our brainy brethren? You can't go visit their likely homes. Rocketing off to distant star systems, although a work-a-day staple of science fiction, is actually quite difficult. The speed of earthly rockets, an impressive 30,000 miles per hour, is less impressive when you reckon that it would take these craft 1,000 centuries to reach Alpha Centauri, the nearest stellar stop on the tour of the universe. Faster rockets take less time, but they consume more energy — a lot more energy.

More than 50 years have passed since astronomer Frank Drake made the first efforts to put us in touch with aliens. So far, our telescopes haven't snagged a single confirmed extraterrestrial peep. But keep in mind that, until now, the search has been limited. As technology (and, one hopes, funding) continues to improve, the chance for success increases. Someday soon, astronomers may begin puzzling over a signal that comes from the cold depths of space. Maybe the signal will teach us interesting lessons, such as the meaning of life, or at least all the laws of physics. But one thing is for sure: The signal will show us that we're not the only kids on the galactic block.

Using Drake's Equation to Discuss SETI

Although Earthlings can't visit distant civilizations, astronomers are trying to find evidence of technically sophisticated space aliens by eavesdropping on their radio traffic. In 1960, astronomer Frank Drake attempted to listen in on cosmic communications by using an 85-foot (in diameter) radio telescope in West Virginia. If you've seen the movie *Contact,* you know that a radio telescope is similar to a seriously bulked-up backyard satellite dish (see Figure 14-1). Drake connected his antenna to a new, sensitive receiver working at 1,420 MHz (located in what's called the microwave region of the radio spectrum) and then pointed the telescope at a couple Sun-like stars.

Drake didn't hear any aliens during his Project Ozma, but he provoked a great deal of enthusiasm within the scientific community. A year later, in 1961, the first major conference on SETI was held. Drake tried to organize the meeting by distilling all the unknowns of the search into a single equation, now known as the *Drake Equation.* (For the mathematically inclined, I provide this simple little formula in the sidebar "Diving into the Drake Equation," later in this chapter.) Its logic is easy. The idea is to estimate N, the number of civilizations in our galaxy that use the radio airwaves now. N clearly depends on the number of suitable stars in the galaxy, multiplied by the fraction that have planets, multiplied by the number of . . . well, you can read about it in the sidebar.

TECHNICAL STUFF

Diving into the Drake Equation

Scientists often use Frank Drake's nifty little formula as the basis for discussions about SETI and the chances that humans will ever make contact with extraterrestrial intelligent life. The equation is quite simple and doesn't require any math beyond what you mastered in the eighth grade.

The equation computes N, the number of broadcasting civilizations active in the Milky Way galaxy. As with the Bible, several versions of Drake's Equation exist, but here's the usual formulation, in all its gory glory:

$N = R^* f_p n_e f_l f_i f_c L$

- R^* is the rate at which long-lived stars, suitable for hosting habitable planets, form in the galaxy. Because the Milky Way has roughly 300 billion stars and is approximately 13 billion years old, this number is about two per year.

- f_p is the fraction of good stars — stars that can have habitable planets — that actually *have* planets. Astronomers don't know what this number is, but recent planet searches indicate that it's at least 50 percent, and maybe much more.

- n_e is the number of planets per solar system that can incubate life. In our solar system, the number is at least one (Earth) and could be more if you count Mars and some of the moons of Jupiter and Saturn. But in another system, who knows? A typical guess is one.

- f_l is the fraction of habitable planets that actually develop life. We can reasonably assume that many of them do.

- f_i is the fraction of planets with life that develop intelligent life. This number is controversial, of course, because intelligence may be a rare accident in biological evolution.

- f_c gives the fraction of intelligent societies that invent technology (in particular, radio transmitters or lasers). Probably most of them do.

- L, the final term, is the lifetime of societies that use technology. This term is a matter of sociology, not astronomy, of course, so your guess is as good as the author's — maybe better.

The number N depends on your choice of values for the various terms. Pessimists think that N may be only 1 (we're alone in the Milky Way galaxy). Carl Sagan figured it to be a few million. And what does Drake himself say? At one time, his guess was "About 10,000." Moderation in all things. You can visit the site www.seti.org to tinker with the Drake Equation by entering your own values and calculating N.

Drake's Equation is truly seductive, and you may want to impress strangers by rattling it off at a dinner party. But although scientists may know or can safely guess at the values of the first few terms in the equation (such as the rate at which stars capable of hosting planets form and the fraction of such stars that actually have planets), we don't have any real knowledge of such details as the fraction of life-bearing planets that can develop intelligent life or the lifetime of technological societies. So Drake's Equation still doesn't have an "answer." But it's a great way to organize the SETI discussion.

Courtesy of Seth Shostak

Figure 14-1:
With the right kind of receiver on a radio telescope, astronomers can listen for signals from other societies.

SETI Projects: Listening for E.T.

Most of the modern SETI efforts follow in Frank Drake's footsteps. In other words, they use large radio telescopes in an attempt to eavesdrop on signals from alien civilizations.

Why use radio? Radio waves move at the speed of light and easily punch through the clouds of gas and dust that fill the space between the stars. In addition, radio receivers can be quite sensitive. The amount of energy required to send a detectable signal from star to star (assuming that aliens wield a transmitting antenna at least a few hundred feet in size) is no more than what your local television station pumps out.

Assuming that researchers do get an interstellar ping, how do they recognize it? They don't look to receive the value of *pi* or some other simple message that proves the aliens have completed junior high. SETI researchers simply look for narrow-band signals.

Narrow-band signals occur at one narrow spot on the radio dial. Only a radio transmitter makes narrow-band emissions. Quasars, pulsars, and even cold hydrogen gas all make radio waves, but their natural static is spread out in frequency — splattered all over the radio spectrum. Narrow-band signals are the mark of transmitters. And transmitters are the mark of intelligence. It takes intelligence (not to mention a soldering iron) to build a transmitter.

Another criterion that SETI researchers insist upon before they can claim to have received a true alien broadcast is that the signal be persistent. In other words, every time they point their telescopes at the source of the signal, they find it. If their meters register only once, the signal is impossible to confirm. They may count it as interference from telecommunication satellites, a software bug, or an ambitious college prank.

In the following sections, I discuss several SETI projects and explain how you can help with the search.

The flight of Project Phoenix

The most sensitive SETI experiment researchers have completed so far is Project Phoenix, run by the SETI Institute in Mountain View, California, from 1995 to 2004. This project was a successor to a NASA SETI program that Congress halted in 1993. Since then, SETI efforts in the United States have been privately funded.

Project Phoenix trained its sights on individual stars, in what is known in the SETI trade as a *targeted search.* Other projects use telescopes to sweep large tracts of the sky. Of course, those broad sweeps let scientists examine more of the heavens, but by concentrating only on nearby, Sun-like stars, a targeted search achieves much greater sensitivity. In other words, it can find far weaker radio signals. Researchers carried out Project Phoenix on a handful of different telescopes, including the 1,000-foot (in diameter) Arecibo radio telescope (in Puerto Rico), the mother of all antennas (see Figure 14-2).

Phoenix (and many other SETI experiments) looked for signals in the microwave region of the radio dial. Aside from their ability to make leftovers palatable, microwaves are the preferred "hailing channel" of the SETI crowd, for two reasons:

Figure 14-2:
A view of the massive Arecibo radio telescope in Puerto Rico that participated in Project Phoenix.

Courtesy of Seth Shostak

✔ The universe is rather quiet at microwave frequencies — you encounter less natural static, a fact that E.T. also knows.

✔ A natural signal generated by hydrogen gas occurs at 1,420 MHz, a frequency located in the microwave region. Because hydrogen is far and away the most abundant element in the cosmos, every alien radio astronomer should be aware of this natural marker — and may be tempted to get our attention (or the attention of any other civilization in space) by sending out a signal near to its frequency on the dial.

But facing facts, scientists really don't know *exactly* where the extraterrestrials may tune their transmitters. To cover as much of the dial as possible, Project Phoenix checked out many millions of channels at once (and, over the course of time, billions of channels for each targeted star).

When Project Phoenix finally stopped its observing program in spring 2004, it had carefully examined about 750 Sun-like star systems. It found no persistent, clearly extraterrestrial signal. But this effort taught researchers how to build an instrument that could, in a few decades, check out a million star systems or more. Those lessons have resulted in the efforts to construct the Allen Telescope Array (see the next section).

Space scanning with other SETI projects

Today several SETI programs dot the astronomy landscape:

- The *Search for Extraterrestrial Radio Emissions from Nearby Developed Intelligent Populations* (nicknamed *SERENDIP*), conducted from the University of California, Berkeley, uses the Arecibo telescope in a "piggyback" mode, collecting data from whatever direction the telescope points. In this way, researchers can use the telescope for SETI even when other astronomers are studying pulsars, quasars, or other natural objects. This apparently aimless approach pays off handsomely in observing time: SERENDIP collects data almost every day, all day.

  You can read the latest news about SERENDIP at the website `http://seti.berkeley.edu/SERENDIP`.

- *Southern SERENDIP* is run by the SETI Australia Centre at the University of Western Sydney. The center uses a 210-foot radio telescope at Parkes, in the sheep and mosquito country a few hundred miles west of Sydney. The southern version is also a piggyback experiment, in that astronomers other than the SETI researchers control the aim of the antenna. Check out the website `http://seti.uws.edu.au/SSERENDIP-1.html`.

- The SETI Institute and the University of California at Berkeley are building a new radio telescope, called the Allen Telescope Array (ATA), designed for efficient SETI searching. The scope eventually will consist of 350 small (6 feet in diameter) antennas spread over a half-mile of California real estate (see Figure 14-3). At press time in 2012, 42 antennas were operational, and observations were underway.

  The Allen Telescope Array's antennas are like synchronized swimmers: They all point in the same direction. But unlike many past SETI instruments, the antennas can observe several candidate star systems simultaneously. Of course, this characteristic speeds up the search, as does the fact that the array can function 24/7 for SETI. Without a doubt, ATA is the most ambitious SETI telescope project to date. You can follow project progress at the ATA website, `www.seti.org/ata`.

In addition to these SETI efforts, a growing number of so-called optical SETI experiments are showing up. Instead of hunting for radio broadcasts, optical projects look for brief but intense flashes of laser light that may come our way from a society keen to get in touch. Optical SETI relies on conventional mirror-and-lens telescopes, outfitted with high-speed electronics to sense and record any alien light bursts. Optical SETI experiments are underway at the University of California at Berkeley, Harvard University, and elsewhere.

Courtesy of the SETI Institute

Figure 14-3:
Upon completion, the Allen Telescope Array will consist of 350 antennas spread across a half-mile in California.

Optical SETI wasn't taken too seriously at first. Some experts thought that searches for alien laser pulses weren't useful. But an English scientist, Stuart Kingsley, kept the faith and conducted observations for years at the Columbus Optical SETI Observatory in Ohio and, more recently, in Bournemouth, U.K.

Although a light signal from a distant world may be obscured by the light from that planet's sun, it's pretty easy to focus laser light with a mirror. Such an optical transmitter can outshine a star — for the billionth of a second that the flash is on! So it makes sense to look for messages sent in this flashy way, and optical SETI researchers have already trained their telescopes on a few thousand nearby stars. So far, no dice.

The SETI Institute offers a free newsletter. If you want to support the search for extraterrestrial broadcasts, you can purchase a membership. See its website at www.seti.org. To download the institute's latest one-hour radio program about SETI and life in space, go to http://radio.seti.org.

Hot targets for SETI

NASA's Kepler satellite, which I discuss later in this chapter, is finding hundreds of planets around stars beyond the Sun. Some of these exoplanets are roughly Earth size, and some are in or near the habitable zones around their stars, where water can remain a liquid on a planetary surface. If there's water, some experts say, life may exist. If there's life on a planet, perhaps it has evolved to an intelligent state. Intelligent life may be interested in sending radio or laser signals to contact folks on other worlds. Such is the theory, anyway, and SETI projects like the ATA are going to target some of these planets, in case E.T. is calling from home.

SETI wants you!

Two major SETI projects seek your participation to help find alien radio signals. The first program just needs some of the unused time on your computer. But the second project puts your brain in the loop, to help recognize suspicious patterns in radio signal data sent direct from the telescope.

SETI@home is a project that divides a huge data stream from ongoing SETI projects among a great many personal computers of folks like you. If you join SETI@home, your computer will automatically process a small chunk of the SETI data when the machine is idle but turned on.

To join the search, visit the website `http://setiathome.ssl.berkeley.edu` and follow the simple instructions to download free software. From then on, when you're not using the computer, it will sometimes light up with a snazzy SETI screensaver. That's the sign that the computer is crunching away on observational data. Now and then, the computer will connect to a server in Berkeley, California, and send the results of its computing.

In SETI@home's first ten years of operation, more than 5 million people worldwide volunteered the use of idle computer time. As their computers reported back to project headquarters, sometimes they revealed that a suspicious signal was found. SETI scientists examined the reports and, up to now, rejected all of them, meaning that they weren't evidence of E.T. But someday E.T. may call, and your computer may identify the ring. I like the tone of that!

When you join the SETILIVE project, live data from the Allen Telescope Array is sent to your computer. You examine the data displays to look for anything suspicious (a tutorial helps you know what to look for). Then you post your report. Now, please don't take offense, but if you report finding E.T., scientists won't stand up and do a victory dance before informing United Nations headquarters. SETI sends the same signals from the ATA to other volunteers. After independent reports from multiple volunteers finger a particular signal, SETI scientists give it the once over. Too much data comes in for the scientists to look at all of it, so the volunteers are the screening committee. Check out the website at `http://setilive.org`.

Discovering Alien Worlds

Exoplanets are planets that orbit stars other than our Sun. At one time, astronomers didn't know whether a single exoplanet existed, but since the 1990s, they've discovered many of them. The known extrasolar planets (as exoplanets are also called) now outnumber the eight planets of our solar system by far.

In this section, I describe how beliefs about exoplanets have changed over time and the ways in which astronomers are now finding them. I introduce you to the main types of exoplanets, with typical examples. And I describe astrobiology, the science that investigates the possibility that life exists on some of these alien worlds.

Changing ideas on exoplanets

Scientists and other scholars wondered for centuries whether there are planets around other stars. Since there was no evidence of them until recently, few believed in exoplanets. Of those who did believe, the most famous was Giordano Bruno, the Italian Renaissance philosopher. He maintained that the stars in the sky resemble our Sun, long before that was known. He theorized that the stars also have planets that are inhabited, like Earth. These were unpopular views, to say the least. Bruno was condemned and burned at the stake in 1600.

Long after Bruno's time, and through most of the 20th century, astronomers doubted that many (or even any) exoplanets exist. They thought that the planets of our solar system must have formed when a passing star nearly collided with the Sun. Supposedly, tidal force exerted by the passing star pulled gas out of the Sun, and some of the gas condensed and formed the planets. But a close encounter of this stellar kind is incredibly rare, since stars are light-years apart. Few, if any, collisions would mean there are few or no exoplanets. Ideas changed in the 1990s, when the Hubble telescope and other new instruments revealed that many newborn stars (the Young Stellar Objects that I describe in Chapter 11) are surrounded by disk-shape clouds of gas and dust. Conditions in these clouds are right for planets to form. In other words, planet birth is a byproduct of common star formation, not rare collisions. But astronomers needed to discover some exoplanets and see if they actually are as common as this theory implies.

Over the years, a few astronomers claimed to find exoplanets. But the reports were disproven or could not be confirmed. Success finally came in 1992, when radio astronomers detected two planets of a pulsar. (I describe pulsars in Chapter 11.) Then in 1995, researchers discovered the first exoplanet orbiting a normal star. Many more astronomers joined the hunt, with newer and more powerful instruments. By February 2012, the searchers had found 760 exoplanets and more than 2,000 other possible planets that need to be confirmed. NASA's Kepler satellite was turning up exoplanets in large numbers, and an expert with the Kepler project estimated that there may be 100 billion planets in our galaxy.

Finding exoplanets

Exoplanets are much dimmer than the stars they orbit, so a planet is almost always lost in the glare of its star. Therefore, astronomers don't look for planets directly. Instead, they seek telltale effects in the appearance of stars that betray their hidden planets.

The most important clues to the existence of an exoplanet are as follows:

- ✔ **A repeated wobble in the motion of a star:** When astronomers find a star wobbling back and forth repeatedly, with each wobble lasting as long as the previous one, they conclude that it has a faint companion. Gravity makes the unseen companion and the visible star follow orbits around their common center of mass, just like the two members of a binary star (see Chapter 11). But if the companion cannot be seen, it may be much smaller and dimmer than a star, and thus a planet. Astronomers have found many such wobbling stars by observing their spectra change with time. The Doppler effect (which I also explain in Chapter 11) reveals the back-and-forth motions.

✔ **Periodic dimming of a star:** When observers make precise measurements of the brightness of a star and detect a very small dimming, it may mean that the star has a planet that is passing across the face of the star, just like a transit of Mercury across the sun (see Chapter 6). The planet blocks a little of the star's light while it passes in front. The fraction of the star's light that's missing during a transit tells astronomers the size of the planet, compared to the star. (The bigger the planet, the larger the drop in brightness.) The interval between two transits is the time it takes the planet to make one orbit around the star: the planet's "year." Kepler and a French satellite called CoRoT find exoplanets by this transit method, and I list a few of these finds and their years in Table 14-1.

✔ **A brief increase in star brightness, followed (or preceded) closely by another fleeting increase:** Some astronomers use telescopes that monitor thousands of stars at once. They're looking for unusual events, when one of the stars brightens noticeably and then fades to its original magnitude over a period of a few weeks. During those weeks, there may be a second brightening lasting just a few hours or days. The brightenings are the result of gravitational microlensing (which I explain in Chapter 11). In this process, the gravity of a dim foreground star magnifies the light of the more distant star for a few weeks, making it appear brighter. When a second, briefer brightening takes place, that additional effect is caused by the gravity of a planet near the foreground star.

Table 14-1			Notable Exoplanets		
Planet	**Mass**	**Size**	**"Year"**	**Distance**	**Description**
PSR 1257+12c	4 M_E	?	67 days	1,600 light-years	Pulsar planet, diamond planet
Kepler-20f	1–2 M_E	1 D_E	20 days	950 light-years	Exo-Earth
Kepler-22b	?	2.4 D_E	290 days	620 light-years	Super-Earth in habitable zone
GJ 1214b	6.6 M_E	2.7 D_E	1.6 days	42 light-years	Super-Earth, water world
Gliese 581d	6 M_E	?	67 days	20 light-years	Rocky super-Earth, possible Goldilocks planet
WASP-12b	1.4 M_J	1.7 D_J	1.1 days	1,400 light-years	Hot Jupiter, likely carbon planet
CoRoT-9b	0.8 M_J	1.1 D_J	95 days	1,500 light-years	Garden-variety Jupiter

Planet	Mass	Size	"Year"	Distance	Description
HIP 13044b	1.3 MJ	?	16 days	2,300 light-years	Intergalactic interloper
Kepler-16b	0.3 M_J	0.8 D_J	229 days	200 light-years	Tatooine planet, orbiting two cool dwarf stars
WASP-17b	0.5 M_J	2.0 D_J	3.7 days	1,300 light-years	Wrongway planet, hot Jupiter

Planet mass is listed in units of the Earth's mass, M_E, or Jupiter's mass, M_J. Size is given in terms of the diameters of Earth or Jupiter, D_E and D_J. A "?" means that the quantity is unknown. The length of a planet's "year," meaning the time it takes to go once around its star, is stated in units of an Earth day, or 24 hours.

The vast majority of exoplanets are found by the wobble, transit, and micro-lensing observations. But two other methods have found small numbers of unusual exoplanets:

✔ **Direct imaging:** In rare cases, astronomers find exoplanets that are *not* lost in the glare of their stars. The planets show up as tiny dots along-side the stars in telescopic photos. Photos taken over a period of time show whether the "dot" moves through space with the star, confirming that it's a planetary companion, not a faint background object. These "direct imaging" planets are younger, larger, and brighter, on average, than most other known exoplanets and are also farther from their stars than the average.

✔ **Pulsar timing:** Radio astronomers measure the arrival times of pulses of radio waves from pulsars, a kind of dead star that I describe in Chapter 11. The pulses come at precisely spaced intervals. But in a few cases, pulses arrive ahead of schedule for awhile, and then behind schedule for an equal period of time, and that pattern keeps repeating. This pattern reveals that the pulsar is going around a small orbit because of the grav-ity of a planet. When the pulsar is on the side of the orbit that's closer to Earth, the pulses arrive early because they have a slightly smaller dis-tance to travel. And when the pulsar is on the far side, pulses come late because they have a longer way to go. Most exoplanets were born along with their stars, but pulsar planets probably formed after their stars died in supernova explosions.

Meeting the (exo)planets

Astronomers probably haven't even found all kinds of exoplanets because some are too small or too rare to show up in current observations. Current knowledge of exoplanets is thus likely incomplete. Yet astronomers have identified many interesting kinds already, most of them unlike any planet in our solar system.

Take a look at the main types of known exoplanets:

- **Carbon planet:** A rocky world with a lot more carbon and much less silicate rock and water (if any) than Earth. Its surface layer may be mostly graphite (like the "lead" in a pencil); underground, the carbon forms a layer of diamond (making it also a *diamond planet*). The interior of a very large carbon planet may hold BC8, a type of carbon with no common name that is even denser than diamond.

- **Exo-Earth (or Earth, for short):** A rocky planet with about the same size and mass as our Earth.

- **Super-Earth:** An exoplanet that's bigger and more massive than an exo-Earth, but smaller than Neptune, more or less. SuperEarths range in mass from twice to ten times the mass of Earth. A super-Earth may be a rocky planet; a gassy, icy planet like Uranus (see Chapter 9); a carbon planet; or even a water world (described later in this list).

- **Goldilocks planet:** An exo-Earth, or a rocky super-Earth, with surface conditions suitable for liquid water to exist. It should be located in the *habitable zone,* which is the range of distances from its star within which surface water neither freezes permanently nor boils away. Some rocky exoplanets with thick atmospheres may be in the habitable zone but are not Goldilocks planets because their atmospheres trap so much heat that the surface temperature is far above the boiling point. (That description sounds just like Venus, which I describe in Chapter 6.)

- **Hot Jupiter (also called a *roaster*):** A gas giant planet like Jupiter (see Chapter 8) that's very close to its sun. Many hot Jupiters are closer to their suns than Mercury is to ours. See Figure 14-4 for an artist's rendering of a hot Jupiter exoplanet.

- **Jupiter:** A gas giant exoplanet that's located far enough from its Sun that it's cold, like our Jupiter. I call this kind of exoplanet a "garden-variety Jupiter," although, so far, astronomers have found far more hot Jupiters than cold ones.

- ✔ **Intergalactic interloper:** An exoplanet that orbits a star in our Milky Way galaxy that has been captured from another galaxy. (As I explain in Chapter 12, the Sagittarius Dwarf Galaxy, for example, is being absorbed by the Milky Way. Gradually, its stars become members of the Milky Way.)

- ✔ **Nomad planet:** A planet-mass body that's located in the space between the stars and that isn't orbiting a star. Some astronomers refer to these exoplanets as "free floaters." Nomads are an exception to the definition of an exoplanet as a body orbiting a star. They may have formed around a star and then somehow escaped from orbit. Or maybe they were born in an as-yet-unknown way.

- ✔ **Tatooine planet:** An exoplanet that orbits a binary star so that it has two suns. The name comes from the fictional planet Tatooine, with two suns and a desert landscape, that's Luke Skywalker's home in the *Star Wars* films.

- ✔ **Tidal Venus:** An exo-Earth that, although it may be located in a habitable zone, is (like Venus itself) much too hot to have liquid water. Our Venus is hot from trapping solar energy in the thick atmosphere. But a tidal Venus is heated by strong tidal force exerted by its sun, which causes friction in its rocky interior.

- ✔ **Water world:** A super-Earth that's substantially made of water. Compare that with our Earth, which is made of rock and iron, with oceans on the surface. A water world can be half or more made of water, with no place to dock your boat. Kevin Costner wouldn't like that.

- ✔ **Wrongway (retrograde) planet:** An exoplanet that orbits its sun in the opposite direction to the way the star turns. In the solar system, all eight planets (and Pluto, too) follow *prograde* orbits. In other words, they orbit in the same way our Sun turns: counterclockwise, as seen from an imaginary point far above the north pole of the Sun. A retrograde planet is like the famous aviator "Wrongway Corrigan," who flew from New York to Ireland in 1938 when he was supposed to head for Long Beach, California.

I list typical examples for most types of exoplanet in Table 14-1. I don't include a tidal Venus because scientists have found no definite case yet. (A few possible tidal Venuses are under study.) And I don't list an intergalactic interloper because, although some have been discovered, the details of individual interlopers haven't been announced.

Many of the known exoplanets are part of larger systems, with two or more planets orbiting the same star, like the eight planets of our Sun. As we go to press, about 130 known exosolar planetary systems exist.

Figure 14-4:
An artist's
concept of a
hot Jupiter
orbiting
the star 51
Pegasi.

Courtesy of Seth Shostak

Checking out planets for fun and science

You can check out the Kepler satellite and its planetary discoveries at the NASA website `http://kepler.nasa.gov`. See the status and findings of France's exoplanet mission, CoRoT, at `http://smsc.cnes.fr/COROT`.

If you have an iPhone or iPad, you can use the free Kepler Explorer app developed by the University of California, Santa Cruz. The app lists the exoplanets that Kepler discovered, shows you how they orbit the host stars, and allows you to select individual exoplanets for further study on your mobile device. Just download the Kepler Explorer from the iTunes App Store.

You can help the scientific research on exoplanets by joining the Citizen Science PlanetHunters project maintained by Yale University and Zooniverse at the website `www.planethunters.org`. Participants take an active role in investigating the data from Kepler.

Astrobiology: How's Life on Other Worlds?

Astrobiology is the science that studies the possibility of life in space. At least some exoplanets must be likely sites for life. That's because, if the galaxy has 100 billion exoplanets, at least a few should have the right conditions for life to arise and flourish. Unfortunately, scientists have found only a tiny fraction of the 100 billion planets that they think exist, and all are very far away. We can't get a closeup picture of an exoplanet with a big telescope and look at large herds of migratory dinosaurs.

The two most fruitful avenues of examining the possibility of extraterrestrial life right now are as follows:

✏ Studying extremophiles, meaning life forms that exist in extreme environmental conditions on Earth — conditions that are deadly to most ordinary kinds of life.

✏ Searching nearby bodies in the solar system with space probes and telescopes for signs of life, current or past.

I describe how researchers use both methods in the following sections.

Extremophiles: Living the hard way

Most extremophiles are microorganisms, like bacteria, although many species of plants survive at temperatures below −40°F in Antarctica. If you enjoy extreme cold, you're a *cryophile*. Some bacteria flourish in water so hot that it would scald to death everything else. These *hyperthermophiles* live at 257°F in hot springs in Yellowstone National Park and at equally high temperatures in deep ocean vents. That temperature is above the normal boiling point, 212°C, but water that hot doesn't boil when under high pressure, deep below the ocean surface. The hyperthermophiles we know of couldn't last on Venus, where the surface temperature is 870°F (see Chapter 6). But there's probably a moderately hot exoplanet somewhere that's just right for organisms like them.

If you're a gardener, you may check the pH of your soil to make sure it's not too alkaline or acidic for the plants you want to grow. Too much alkali (or acid) can be fatal for most living things. But certain extremeophiles love ultra-alkaline conditions *(alkalophiles)* or swim through acid waters with impunity *(acidophiles)*. Drop a fish in the Dead Sea, and it dies (hence, the name). But for some bacteria *(halophiles)*, that salty sea is sweet indeed.

Some bacteria live in tiny pores in solid rock more than 3 miles underground. Those extremophiles don't get energy from the Sun; they run on chemical energy from their immediate environment. And *barophiles* thrive in the ocean depths, where the pressure of the overlying water is a thousand times the atmospheric pressure at sea level. Scientists have even discovered bacteria that live in the clouds above your head, and I'm not counting the ones that are just lofted up from the ground by the wind.

What extremophiles tell us is that life is opportunistic. It will find ways to survive — and perhaps to arise in the first place — in conditions under which we can't exist. What can happen on Earth can happen on exoplanets. And that's just assuming that life on other planets is like life as we know it. If other life is completely unlike what we know — for example, life that's not based on carbon (like Earth life), but on another element — all bets are off, and many otherworldly environments may be inhabited.

Seeking life in the solar system

Obtaining reliable evidence of life off Earth is going to be difficult. But we'll never know if we don't try. Astronomers have identified the following locations as the best places to look for life in the solar system (outside Earth):

- ✔ Mars
- ✔ Jupiter's moon Europa
- ✔ Saturn's moons Titan and Enceladus

The ongoing search for Martians

As I describe in Chapter 6, scientists examined and disputed the claim that microscopic fossils are present in a rock from Mars. Astronomers studying Mars with telescopes on Earth get conflicting results about whether methane gas appears from time to time on Mars. One source of methane could be bacteria (some species on Earth make methane, and others eat it). But even if methane does exist on Mars, it may come from a geological process, just as some gases on Earth come out of volcanoes. NASA space probes mapped Mars landforms that look like the remains of ancient sea beds and flood channels. Robotic Mars rovers have found rocks and minerals that may have formed when pools of water dried up. Where there was water on Mars, there may have been life. Some scientists believe some parts of Mars have permafrost. Below the permafrost, where it's warmer underground, there could be liquid water. Perhaps microbes are present on Mars now, living beneath the reach of the rovers' scoops. Inquiring minds want the scoop.

The latest NASA Mars mission, the Mars Science Laboratory, reached the red planet after we went to press. It carries a robot rover, *Curiosity*. The rover's chief purpose is to verify whether past conditions on Mars could have supported microbial life (or whether present conditions are suitable to do so). You can follow its progress through space and on Mars at `http://mars.jpl.nasa.gov/msl`.

Europa, we hardly know thee

Europa is a rocky Jupiter moon with a surface layer of ice (see Chapter 8). Liquid water (probably salty) covers the whole moon under the ice. Scientists believe the ocean could host microscopic life. However, the ice is at least a mile or two thick, and possibly much thicker. We may not have the engineering capability to drill down to the ocean on Europa anytime soon, even if NASA or another space agency had the money to try it. Some Europa experts suspect that there's occasional turnover in the ice and ocean. If so, some water gets to the surface, where it freezes and remains. If that's true, the upheavals may bring traces of ocean life to the surface, where a lander probe may find it.

Titan: Perhaps a primeval Earth

Titan, the largest moon of Saturn, is bigger than Earth's Moon and has a thick atmosphere, making it more like a planet than any other moon (see Chapter 8). It also has large lakes of liquid hydrocarbons, which are the only bodies of liquid on the surface of any object in the solar system besides Earth. Astronomers think Titan may resemble the very early Earth, before oxygen became an important part of the atmosphere. So if life arose on Earth, it could happen on Titan. Titan is colder than Earth ever was, but it does have an underground water ocean. (For more on Titan, check out Chapter 8.)

The Enceladus plume

Enceladus is an icy moon of Saturn that has a body of water beneath the ice, at least near the south pole. Quantities of water that freezes immediately into tiny ice particles spout out from the polar region into space. Unlike on Europa, where the ocean is deep beneath thick ice, Enceladus's water is close to the surface. Because of the water's proximity to the surface, scientists may find it easier to sample and look for evidence of life.

Mooning over extraterrestrial life

The fact that moons like Europa, Titan, and Enceladus may be capable of supporting life has an important implication for astrobiology: When astronomers think about life on exoplanets, they need to consider the moons as well. Astronomers are starting to search for exomoons because where there are exoplanets, there should be some moons, too. If a planet is in a habitable zone but is itself unsuitable for life (because its atmosphere traps too much heat, for example), it may have a moon that's just right for life in the zone.

Dr. Seth Shostak, senior astronomer at the SETI Institute in Mountain View, California, contributed this chapter to earlier editions of Astronomy For Dummies. *The author, Stephen P. Maran, revised the section on SETI and wrote the section on exoplanets for this edition.*

Chapter 15

Delving into Dark Matter and Antimatter

S tars and galaxies set the night sky aglow, but these glittering jewels account for only a tiny portion of the matter in the cosmos. There's more to the universe than meets the eye — much more.

This chapter introduces you to the concept of dark matter, tells you why astronomers are convinced that the stuff exists, and describes experiments that may shed light on the nature of this mysterious, invisible material. I also discuss another exotic type of matter in the universe: antimatter. Yes, antimatter exists outside of science fiction, and the real-world version is every bit as fascinating as the sci-fi books, television shows, and movies suggest.

Dark Matter: Understanding the Universal Glue

As far back as the 1930s, an astronomer found hints that the vast majority of the mass in the universe doesn't emit, reflect, or absorb light.

The invisible material, known as *dark matter,* serves as the gravitational glue that keeps a rapidly rotating galaxy from flying apart and enables fast-moving galaxies in a cluster to stick together. Dark matter also seems to have played a crucial role in the development of the universe as we know it today — a spidery network of immensely long superclusters of galaxies separated by giant voids (see Chapter 12).

Astronomers have determined that almost 84 percent of all the matter in the universe is dark matter. What a humbling thought. The universe you see when you peer through a telescope or look up at the night sky teeming with stars and galaxies is just a tiny fraction of what's out there. To borrow a nautical analogy, if galaxies are like sea foam, dark matter is the vast, unseen ocean on which it floats.

Gathering the evidence for dark matter

The first hint that the universe contains dark matter appeared in 1933. While examining the motions of galaxies within a large cluster of galaxies in the constellation Coma Berenices, astronomer Fritz Zwicky of the California Institute of Technology found that some galaxies move at an unusually high speed. In fact, the galaxies of the Coma Cluster move so rapidly that all the visible stars and gas in the cluster can't possibly keep the galaxies gravitationally bound to one another, according to our known laws of physics. Yet somehow the cluster remains intact. (The Coma Cluster is about 320 million light-years from Earth. To learn more about clusters of galaxies, check out Chapter 13.)

Zwicky concluded that some sort of unseen matter must exist within the Coma Cluster to provide the missing gravitational attraction.

Scientists often fail to appreciate an astonishing discovery when just one person or team claims it. They want to see more evidence, from independent experts, before they are convinced of the new findings. So it's not surprising that dark matter didn't make headlines for several decades after Zwicky's work. Many astronomers ignored his report or figured that after they studied the motions of galaxies in greater detail, the rationale for the existence of the supposedly invisible material would disappear.

In the 1970s, astronomers began to find additional and compelling evidence for dark matter. Not only do clusters of galaxies seem to contain the dark stuff, but dark matter is also in individual galaxies. The following sections describe the main arguments in favor of dark matter.

Dark matter makes stars orbit oddly

Vera Rubin and Kent Ford of the Carnegie Institution of Washington, D.C., were studying the motions of stars in hundreds of spiral galaxies when they obtained a result that seemed to fly in the face of conventional physics. A typical spiral galaxy resembles a flattened fried egg, with most of its visible mass (stars and bright nebulae) seemingly concentrated in the yolk — astronomers call this the *bulge* (as I explain in Chapter 12). Images reveal that the visible mass of a spiral diminishes rapidly with increasing distance from the bulge.

Astronomers presumed that the stars in a spiral galaxy orbit around the galaxy center in the same way that the planets in our solar system orbit the Sun. Obeying Newton's law of gravity, the outer planets, such as Uranus and Neptune, orbit the Sun less rapidly than the inner planets, such as Mercury and Venus. Therefore, stars on the outskirts of a spiral galaxy should orbit at a lower speed than stars nearer the center. But that's not what Rubin and Ford found.

In galaxy after galaxy, their observations revealed that the outlying stars orbit rapidly, just like the inner stars. With so little visible material in the outer regions, how do the outlying stars manage to zip around so fast and still stay bound to the galaxy? They should escape from their galaxy, given their speed. Rubin had found serious hints of this unexpected behavior in earlier work, but many astronomers needed to be convinced of it. (For more about escape velocity, see Chapter 13.)

After the findings of Rubin and Ford became well known, astronomers concluded that *visible matter* — the stars and luminous gas that show up on telescopic photographs — makes up only a small portion of the total mass of a spiral galaxy.

Although the visible mass is indeed concentrated toward the center, a vast quantity of other material must extend far beyond. Each spiral galaxy is surrounded by an immense *halo* of dark matter. And to exert enough of a gravitational tug on the stars in the visible outskirts of the galaxies to make them orbit as rapidly as observed, the dark matter must exceed the visible matter by a factor of 10 in mass. Other types of galaxies, including elliptical and irregular galaxies, also have dark-matter halos. Dwarf galaxies (which I describe in Chapter 12) have even higher proportions of dark matter to visible matter than large galaxies.

Starting in the 1990s, astronomers who observed clusters of galaxies with X-ray telescopes on satellites such as ROSAT and the Chandra X-ray Observatory mapped huge regions in and around the clusters that glow in X-rays. The glows come from an *intracluster medium* of thin, very hot gas. The hot gas fills such a huge region that, thin as it is, its total mass exceeds the sum of the masses of all the galaxies in a cluster.

The thin, hot gas would expand away, but it's kept in place at the cluster of galaxies by the gravitational attraction of a mass that is much larger than its own mass plus the mass of the cluster galaxies. The source of the powerful gravity is dark matter in the cluster. This fact is further evidence that Fritz Zwicky was right in 1933 when he claimed that unseen matter (which we now call dark matter) exists in huge amounts in a cluster.

Cold dark matter makes the universe take its lumps

Cosmologists (scientists who study the large-scale structure of the universe and its formation) also point to dark matter to explain a fundamental puzzle about the universe: How did it evolve from a nearly uniform soup of elementary particles in the aftermath of the Big Bang (see Chapter 16) to reach its present lumpy structure of galaxy clusters and superclusters?

Even though 13.7 billion years have passed since the birth of the universe, scientists don't believe that enough time has elapsed for visible matter to coalesce on its own into the huge cosmic structures we see today.

To solve this cosmological conundrum, some experts state that the universe contains a special type of dark matter, called *cold dark matter,* that moves more slowly but gathers into clumps more quickly than ordinary, visible matter. Responding to the tug of this exotic material, the ordinary matter formed stars and galaxies within the densest concentrations of the dark matter. This theory explains why every visible galaxy seems to be embedded in its own dark-matter halo.

Is the cold dark matter theory correct? It appears to be in broad agreement with the facts about the universe, as far as scientists know them. But the agreement isn't perfect. For example, the theory predicts that hundreds of tiny satellite galaxies will surround a big galaxy like the Milky Way. But we don't see all that many satellite galaxies. The predictions of the theory may need work, or we may need a better theory of dark matter itself. Or maybe there are small, faint galaxies all around us that we haven't found yet.

It's even possible that astronomers haven't found enough satellite galaxies because many of them were swallowed by the large galaxies that they orbit, like the Sagittarius Dwarf Galaxy and the Canis Major Dwarf Galaxy, which are being absorbed now by the Milky Way (as I tell in Chapter 12).

Dark matter is critical to the universal density

Astronomers believe in dark matter for yet another cosmic reason: On large scales, the universe looks the same in all directions and has an overall smoothness. This consistency in appearance and smoothness indicates that the universe has just the right density of matter, called the *critical density* (which I explain in Chapter 16). The total amount of visible matter that we observe in the universe isn't nearly enough to achieve critical density. Dark matter takes up the slack.

Debating the makeup of dark matter

Okay, astronomers have plenty of good reasons to believe in dark matter. But what the heck is the stuff anyway?

Broadly speaking, astronomers divide the possible kinds of dark matter into two classes: baryonic dark matter and oddball dark matter.

Baryonic dark matter: Lumps in space

Some dark matter may consist of the same stuff that the Sun, planets, and people are made of. This kind of dark matter would be part of the family of *baryons,* a class of elementary particles that includes the protons and neutrons found in the nuclei of atoms. But it's not the cold dark matter that I mention in the preceding section.

Baryonic dark matter includes all hard-to-see material that's made of known kinds of matter, including asteroids, brown dwarfs, and white dwarfs (I describe the dwarfs in Chapter 11). Yes, scientists can detect asteroids in our solar system and white dwarfs and brown dwarfs nearby in the Milky Way. But far out in the galactic halo, these objects may be undetectable with present equipment. Such hypothetical objects in the galactic halo, referred to as *MACHOs (massive compact halo objects),* may account for the dark-matter halos surrounding individual galaxies. (I cover the search for MACHOs later in this chapter.) But we don't see nearly enough of them to account for the development of large-scale structure in the cosmos. I think this theory is probably wrong, and the scientists who proposed it will have to take their lumps.

Oddball dark matter: Stranger still

Alternatively, dark matter may consist of one or more types of exotic subatomic particles that bear little or no resemblance to baryons. These particles include *neutrinos,* which do exist (see Chapter 10 for more about them), and others with names such as *axions, squarks, photinos,* and *neutralinos* that physicists have dreamt up without proof of their existence. Experiments are underway, but no one has captured an axion or any other hypothetical dark matter particle thus far — at least, they haven't done it to the satisfaction of other scientists, so we are still in the dark about dark matter.

During the *Big Bang* at the birth of the universe (see Chapter 16), a zoo of weird dark matter particles may have formed, and some types of them may still exist. The theorized particles include the axion, a kind of miniature black hole 100 billion times lighter than an electron. Even though axions (if real) are featherweights, enough of them can contribute significantly to the cosmic mass. Recent experiments suggest that the neutrino (a particle that scientists once thought may have zero mass) does have a very small but real mass. So neutrinos may account for a small portion of the dark matter.

Other candidates for oddball dark matter are heavier — about ten times the mass of the proton — but still insubstantial in terms of making up the dark matter in the universe, unless they occur in large numbers. These include the yet-to-be-detected partners of certain known subatomic particles such as quarks and photons. These hypothetical dark matter counterparts are squarks and photinos. Many theories exist for these dark matter particles, along with fanciful names for them. Scientists describe them collectively as *weakly interacting massive particles,* or *WIMPs* (covered later in this chapter).

Taking a Shot in the Dark: Searching for Dark Matter

Around the world, physicists are planning or operating sensitive detectors to find the elusive, telltale signals of dark matter. Some of these detectors are designed to analyze the subatomic debris created by giant atom-smashing devices, which briefly reproduce the extreme heat, energy, and densities that were present in the early universe.

The search techniques have to be innovative. After all, scientists are hunting material that, by definition, we can't see and, aside from exerting a gravitational force, that doesn't interact with other matter.

All methods for detecting and measuring dark matter are indirect, but attempting to understand dark matter isn't a trivial pursuit. As the dominant form of matter in the universe, dark matter profoundly influenced the past development of the universe and will affect its future as well.

Looking for WIMPs and other microscopic dark matter

Astronomers discovered dark matter through their observations of galaxies. Now physicists are conducting experiments to find dark matter particles and learn what they are.

Here are some dark matter experiments that are currently underway:

- Large particle detectors lie in deep underground laboratories, where the surrounding rock reduces interference from cosmic rays (high-speed, electrified subatomic particles of known types that come from space). As Earth moves through the dark matter of the Milky Way, dark matter particles may be found when they strike the detectors.

- Telescopes are equipped to detect celestial gamma rays, on spacecraft such as NASA's Fermi Gamma-ray Space Telescope. An unusual feature in the spectrum of gamma rays may indicate that some of them come from dark matter particles annihilating in space.

- Arrays of telescopes on Earth detect flashes of visible light that are triggered when celestial gamma rays strike the atmosphere. A prime example is the High Energy Stereoscopic System (HESS) in Namibia. Analysis of these data, like the Fermi data, can reveal a spectral feature of the incoming gamma rays that is traceable to dark matter at their source.

- The Alpha Magnetic Spectrometer (AMS-02) operates on the International Space Station. AMS is searching for unusual cosmic rays that may be produced by *neutralinos,* a certain type of theorized dark matter particles. When neutralinos collide with each other in space, they may create the cosmic rays that AMS is looking for.

Advances in technology are making other experiments concerning dark matter possible:

- Powerful atom smashers, such as the Large Hadron Collider (LHC), near Geneva, Switzerland, can make subatomic particles knock into each other at very high energies. They can generate dark matter particles for detection in the laboratory.

✔ Underground neutrino observatories (see Chapter 10) can be upgraded to make improved measurements of neutrinos from the Sun, to reveal physical conditions near the solar center. These experiments will test the theory that the Sun has accumulated a great deal of dark matter, concentrated toward its center.

For more information on some of these dark matter experiments, consult the following sites:

✔ **NASA's Fermi Gamma-ray Space Telescope website (**`www.nasa.gov/ mission_pages/GLAST/main/index.html`**):** Find news of the dark matter search and other Fermi findings.

✔ **The Alpha Magnetic Spectrometer website (**`http://ams.nasa.gov`**):** A counter tells you how many billion cosmic rays it has measured since it was installed on the International Space Station in May 2011.

✔ **The High Energy Stereoscopic System website (**`www.mpi-hd.mpg.de/ hfm/HESS`**):** Read up on the HESS's many discoveries about astronomical sources of gamma rays.

✔ **Large Hadron Collider (**`http://public.web.cern.ch/public/en/ lhc/lhc-en.html`**):** This site is the public website of the LHC, operated by CERN, the European Organization for Nuclear Research.

MACHOs: Making a brighter image

Because MACHOs aren't microscopic like WIMPS, looking for them is easier. The prime method takes advantage of a mind-bending concept from Einstein's Theory of General Relativity. To wit: Mass distorts the fabric of space and the path of a light wave (as I describe in Chapter 11), which means that an object that happens to lie along the line of sight between Earth and a distant star focuses the light from that star, briefly making it appear brighter. The more massive the object — in this case, a MACHO — the brighter the star appears during the alignment.

In effect, the MACHO acts as a miniature gravitational lens, or microlens, bending and brightening the light from the background star. (See Chapter 11 for more on microlensing.)

To search for MACHOs, astronomers have monitored the brightness of stars from one of the Milky Way's nearest neighbors, the Large Magellanic Cloud galaxy. To reach Earth, starlight from the Cloud must pass through the halo of the Milky Way, and MACHOs that reside in the halo should have a measurable effect on that light.

Astronomers have recorded several events in which stars from the Large Magellanic Cloud suddenly brightened and then dimmed again. The number of MACHOs deduced from these observations is nothing much to phone home about, however. Sorry, E.T.

Mapping dark matter with gravitational lensing

On much larger scales, scientists are taking advantage of gravitational lensing to map the dark matter in entire galaxies and even clusters of galaxies.

If a cluster happens to lie in the travel path of light from a background galaxy, it bends and distorts the light — a process called *gravitational lensing* — creating multiple images of the background body. A halo of these ghost images forms in and around the cluster, as seen from Earth.

Astronomers have used the Hubble Space Telescope to photograph some clusters of galaxies where large numbers of the ghost images of a more distant galaxy appear as short, bright arcs seen against a cluster.

To create the exact pattern of observed ghost images, the intervening cluster must have its mass distributed in a particular way. Because most of the cluster's mass consists of dark matter, the process of gravitational lensing reveals how the dark matter is concentrated in the cluster.

Dueling Antimatter: Proving That Opposites Attract

Get ready for another type of matter almost as weird as dark matter — or maybe even weirder. I'm talking about antimatter.

British physicist Paul Dirac predicted the existence of antimatter in 1929. He combined the theories of quantum mechanics, electromagnetism, and relativity in an elegant set of mathematical equations. (If you want to know more about his theories, you'll have to look them up; this isn't a physics book.)

Dirac found that, for every subatomic particle, a mirror-image twin should exist, identical in mass but with an opposite electrical charge. Thus, the proton has its antiproton and the electron its antielectron.

When a particle and its antiparticle meet, they annihilate each other. Their electric charges cancel out, and their mass is converted into pure energy.

Astronomers have detected antiparticles of the electron and proton in the cosmic rays from deep space. The antielectron is called the *positron,* and the antiproton is simply the *antiproton.* The Alpha Magnetic Spectrometer (AMS-02) on the International Space Station, which I mention earlier in this chapter, is searching for antihelium that also may exist in cosmic rays. Physicists have actually made antiparticles and even entire antiatoms, such as antihydrogen, in the laboratory. Doctors use beams of antiparticles to diagnose and treat cancer.

Physicists have even found a belt of antiprotons within Earth's Van Allen radiation belts (which I describe in Chapter 5). Researchers from Italy and elsewhere discovered the antiproton belt in 2011 with the PAMELA (Payload for Antimatter Matter Exploration and Light-nuclei Astrophysics) particle detector experiment on Russia's Resurs DK1 satellite. (For more on PAMELA and its findings, check out the project website, at `http://pamela.roma2.infn.it/index.php`.)

Astronomers studying high-energy radiation from space have observed a type of gamma ray known as *annihilation radiation.* When an electron and its antiparticle, the positron, meet, they annihilate, releasing gamma rays at a known energy of 511 kiloelectron volts (keV). These telltale rays have been detected from several places in the galaxy, including a wide region in the direction toward the center of the Milky Way. (You can see a map of the annihilation radiation from the Milky Way, as measured by the European Space Agency's INTEGRAL satellite, at `http://sci.esa.int/science-e/www/object/index.cfm?fobjectid=45328`.)

Annihilation radiation has also been detected from some powerful solar flares (see Chapter 10 for more about solar flares).

On the cosmic scale, the big mystery is why the universe contains many more particles than antiparticles. Experiments are underway to find the answer. Many physicists think that the Big Bang forged equal numbers of both. If that's what happened, an unknown physical process has since altered that original balance between matter and antimatter in favor of matter.

Another theory, which many astronomers doubt, is that most of the antimatter from the Big Bang has been corralled into separate regions in distant parts of the universe, where it is in the majority. If people like us, who are made of ordinary matter, went to those regions, we would be destroyed as antimatter annihilated our matter.

At least we know that we have billions of years to solve the problem of the relative lack of antimatter before the universe (and us with it!) slips away to its ultimate fate (which I discuss in the next chapter).

Ron Cowen, who writes about astronomy and space for many publications, originally contributed this chapter. The author, Stephen P. Maran, updated it for the second edition and this edition of Astronomy For Dummies. *All opinions expressed in this chapter are those of the author.*

Chapter 16

The Big Bang and the Evolution of the Universe

*O*nce upon a time, 13.7 billion years ago, the universe as we know it didn't exist. No matter, no atoms, no light, no photons — not even space or time.

Suddenly, perhaps in an instant, the universe took form as a tiny, dense speck filled with light. In a minuscule fraction of a second, all the matter and energy in the cosmos came into being. Much smaller than an atom, the infant universe was searingly hot, a fireball that began mushrooming in size and cooling at a furious rate.

Astronomers and people the world over have come to know this picture of the birth of the universe as the *Big Bang* theory.

The Big Bang wasn't like a bomb that explodes into the environment — there was no environment until the Big Bang occurred. It was the origin and rapid expansion of space itself. During the first trillion-trillion-trillionth of a second, the universe grew more than a trillion-trillion-trillion times bigger. From an original smooth mixture of subatomic particles and radiation arose the collection of galaxies, galaxy clusters, and superclusters present in the universe today. It boggles the mind to think that the largest structures in the universe, congregations of galaxies that stretch hundreds of millions of light-years across the sky, began as subatomic fluctuations in the energy of the infant cosmos. But that's what scientists believe about how the universe took shape.

In this chapter, I cover evidence supporting the Big Bang theory, the expansion of the universe, and related information on dark energy, the cosmic microwave background, the Hubble constant, and standard candles.

For more information on the concepts in this chapter, visit UCLA's Frequently Asked Questions in Cosmology site, at www.astro.ucla.edu/~wright/cosmology_faq.html. The site is maintained by Professor Ned Wright, who gives you the right stuff.

Evidence for the Big Bang

Why believe that the universe began with a bang?

Astronomers cite three different discoveries that make a compelling case for the theory:

- ✔ **The expanding universe:** Perhaps the most convincing evidence for the Big Bang comes from a remarkable discovery made by Edwin Hubble in 1929. Up to that time, most scientists viewed the universe as static — stock still and unchanging. But Hubble discovered that the universe is expanding. Groups of galaxies are flying away from each other, like debris flung in all directions from a cosmic explosion, but they haven't been flung apart into space; the space itself between them is expanding, which makes them move farther and farther apart.

    It stands to reason that if galaxies are flying apart, they were once closer together. Tracing the expansion of the universe back in time, astronomers — aided by telescopes and by observatories in space — found that, 13.7 billion years ago (give or take 100 million years), the universe was an incredibly hot, dense place in which a tremendous release of energy triggered an enormous explosion.

- ✔ **The cosmic microwave background:** In the 1940s, physicist George Gamow realized that a Big Bang would produce intense radiation. His colleagues suggested that remnants of this radiation, cooled by the expansion of the universe, may still exist — like the fumes that persist from an extinguished house fire.

    In 1964, Arno Penzias and Robert Wilson of Bell Laboratories were scanning the sky with a radio receiver when they detected a faint, uniform crackling. What the researchers first assumed was static in their receiver turned out to be the faint whisper of radiation left over from the Big Bang. The radiation is a uniform glow of microwave radiation (short radio waves) permeating space. This *cosmic microwave background* has exactly the temperature that astronomers calculate it should (2.73 K above absolute zero, which is −273.16°C or −459.69°F) if it has cooled steadily since the Big Bang. For their historic discovery, Penzias and Wilson shared the 1978 Nobel Prize in physics. (See the section "Universal Info Pulled from the Cosmic Microwave Background," later in this chapter, for the full scoop.)

✔ **The cosmic abundance of helium:** Astronomers have found that the amount of helium among all the baryonic matter in the universe is 24 percent by mass (the rest of baryonic matter is almost entirely hydrogen; iron, carbon, oxygen, and all that good stuff put together is just a trace constituent, compared to hydrogen and helium). Nuclear reactions inside stars (see Chapter 11) haven't gone on long enough to produce this amount of helium. But the helium we've detected is just the amount that the theory predicts would've been forged in the Big Bang. Adding to the evidence, NASA's Wilkinson Microwave Anisotropy Probe (WMAP) found that there was helium in the early universe before there were stars.

Astronomers have other observational evidence that the universe has expanded and changed with time besides the three discoveries I just listed. For example, the deepest photographs of space made with the Hubble Space Telescope reveal that galaxies in the early universe were often smaller or more irregular in shape and more likely to be colliding with each other than present-day galaxies. That information is consistent with the idea that the universe was much smaller back then, so galaxies were closer together and more likely to collide. If the universe had recently formed from a Big Bang, the galaxies themselves would have been younger and smaller.

Such evidence makes clear that the universe is evolving in ways that are consistent with the idea that it began with the Big Bang and has gotten bigger over time.

As successful as the standard Big Bang theory has proved to be in accounting for observations of the cosmos, the theory is but a starting point for exploring the early universe. For example, despite its name, the theory doesn't suggest a source for the cosmic dynamite that sparked the Big Bang in the first place.

Inflation: A Swell Time in the Universe

Aside from ignoring the source of the expansion-causing explosion, the Big Bang theory has other shortcomings. In particular, it doesn't explain why regions of the universe that are separated by distances so vast that they can't communicate — even by a messenger traveling at the speed of light — look so similar to each other.

In 1980, physicist Alan Guth devised a theory, which he called *inflation,* that can help explain this puzzle. He suggested that a tiny fraction of a second after the Big Bang, the universe underwent a tremendous growth spurt. In just 10^{-32} seconds (a hundred-millionth of a trillionth of a trillionth of a second), the universe expanded at a rate far greater than at any time in the 13.7 billion years that have elapsed since.

This period of enormous expansion spread tiny regions — which had once been in close contact — to the far corners of the universe. As a result, the cosmos looks the same on the large scale, no matter what direction you point a telescope. (Think of a big, lumpy ball of dough: If you roll the dough over and over again with your rolling pin, eventually you smooth out all the lumps and create a uniform sheet of dough.) Indeed, inflation expanded tiny regions of space into volumes far bigger than astronomers can ever observe. This expansion suggests the intriguing possibility that inflation created universes far beyond the scope of our own. Instead of a single universe, a collection of universes, or a *multiverse,* may exist. But I'm adverse to that theory — one universe is hard enough to comprehend as it is!

Inflation had another effect: The infinitesimally short but extraordinarily great growth spurt after the Big Bang captured random, subatomic fluctuations in energy and blew them up to macroscopic proportions. By preserving and amplifying these so-called *quantum fluctuations,* inflation produced regions of the universe with slight variations in density from one to another.

Because of inflation and the quantum fluctuations, some regions of the universe contain more matter and energy, on average, than other regions. As a result, there are cold spots and hot spots in the temperature of the cosmic microwave background (see Figure 16-1). Over time, gravity molded these variations into the spidery networks of galaxy clusters and giant voids that fill our universe today, as I describe in Chapter 12. Check out "Universal Info Pulled from the Cosmic Microwave Background," later in this chapter, for more information.

Figure 16-1:
A "baby pic-
ture" of the
universe,
from the
Wilkinson
Microwave
Anisotropy
Probe
satellite.

Courtesy of NASA/WMAP Science Team

The following sections cover two other interesting facets of inflation: the vacuum where inflation gets its power and the relationship between inflation and the universe's shape.

Something from nothing: Inflation and the vacuum

Ironically, the reservoir of energy that powers inflation comes from nothing: the *vacuum*. According to quantum theory, the vacuum of space is far from empty. It seethes with particles and antiparticles that are constantly being created and destroyed. Tapping into this energy, theorists suggest, provided the Big Bang with its explosive energy and radiation.

The vacuum has another bizarre property: It can exert a repulsive force. Instead of pulling two objects together as gravity does, this force pushes them farther apart. The repulsive force of the vacuum may have led to the brief but powerful era of inflation.

Just as in the economy, cosmic inflation generates high interest. And this bubble won't burst.

Falling flat: Inflation and the shape of the universe

The inflation process — at least, in its simplest imagined form — would have imposed another condition on the universe: making the geometry of the universe flat. This rapid period of expansion would have stretched out any curvature in the cosmos like a balloon blown to enormous proportions.

For the universe to be flat, it must contain a very specific density, called the *critical density.* If the density of the universe is greater than the critical value, gravity's pull will be strong enough to reverse the expansion, eventually causing the universe to collapse into what astronomers call the *Big Crunch.*

Such a universe would curve back on itself to form a closed space of finite volume, like the surface of a sphere. A starship traveling in a straight line would eventually find itself back where it started. Mathematicians call this geometry *positive curvature.*

If the density is less than the critical value, gravity can never overpower the expansion, and the universe will continue to grow forever. Such a universe has *negative curvature,* with a shape akin to a horse's saddle.

Although inflation theory demands that the universe is flat, several types of observations have revealed that the universe doesn't have enough matter (whether normal or dark matter — see Chapter 15) to attain critical density.

So if the universe is flat, matter as we know it — or even as we don't know it — can't do the trick. But like Mighty Mouse, energy can save the day! In fact, it can save the universe, and recent research shows that it does. The data contained in the "baby picture of the universe" shown in Figure 16-1, which is a sky map of the cosmic microwave background radiation as measured by the WMAP satellite, has convinced essentially all cosmologists that the universe is flat and that energy is responsible. But it's not energy as we ever knew it before; dark energy is the hero. Read on to discover the dark side.

Dark Energy: The Universal Accelerator

Dark energy has a startling effect: It exerts a repulsive force throughout the universe. That's all scientists know about it. We don't know what dark energy is, so we define it by its observable property, the repulsive force. After the Big Bang and inflation occurred, gravity slowed the expansion of the universe. But as the universe grew bigger and bigger so that the matter was spread apart into more and more space, the slowing effect of gravity became weaker. After a while (some billions of years), dark energy's repulsive force took over, causing the universe to expand ever faster. Observations from the Hubble and other telescopes have revealed this bizarre phenomenon.

The observations that revealed the existence of dark energy — by showing that the expansion of the universe is speeding up — were of Type 1a supernovas in distant galaxies. (You can read about Type Ia and other supernovas in Chapter 11.) All supernovas are bright enough to be seen in distant galaxies, but the Ia supernovas have a special property. Astronomers believe that these explosions all have roughly the same intrinsic brightness, like incandescent light bulbs of a particular wattage (see the section "In a Galaxy Far Away: Standard Candles and the Hubble Constant," later in this chapter).

Because light from a distant galaxy takes hundreds of millions of years or more to reach Earth, observations of that galaxy may show supernovas that erupted when the universe was much younger. If the expansion of the universe had been slowing ever since the Big Bang, there would be less distance between Earth and the faraway galaxy — and a shorter travel time for light — than if the universe had continued to expand at a fixed speed. Thus, in the case of a slower expansion, a supernova from a distant galaxy should look slightly brighter.

But in 1998, two teams of astronomers found exactly the opposite result: Distant supernovas looked slightly dimmer than expected, as if their home galaxies were farther away than calculated. It appears that the universe has revved up its rate of expansion. This discovery revealed the presence of dark energy, as I describe in Chapter 11, and earned the Nobel Prize in Physics for three astronomers — Saul Perlmutter, Adam Riess, and Brian Schmidt — who led the research.

Universal Info Pulled from the Cosmic Microwave Background

The cosmic microwave background (the faint whisper of radiation left over from the Big Bang) represents a snapshot of the universe when it was 379,000 years old. Before that time, a fog of electrons pervaded the infant universe, and radiation created in the Big Bang couldn't stream freely through space. The negatively charged particles repeatedly absorbed and scattered the radiation.

Around the time that the cosmos celebrated its 379,000th birthday, the universe became cool enough for electrons to combine with atomic nuclei, meaning there wasn't an abundance of particles to scatter and absorb radiation. The absorbing fog was lifted. Today we detect the light from the universe at age 379,000 years — now shifted in wavelength by the expansion of the universe — as microwaves and far-infrared light.

Finding the lumps in the cosmic microwave background

When Penzias and Wilson first detected the cosmic microwave background in the 1960s, it appeared to have a perfectly uniform temperature across the sky. No regions in the sky were ever so slightly hotter or colder — at least, not to the detection limits of the available instruments. That uniformity was a puzzle because such tiny variations in temperature have to be present to explain how the universe could have begun as a smooth soup of particles and radiation and evolved into a lumpy collection of galaxies, stars, and planets.

According to theory, the infant universe wasn't perfectly smooth. Like lumps in a bowl of porridge, it had slightly overdense and underdense places, with more atoms per cubic inch or fewer atoms per cubic inch, respectively. These places represent the tiny seeds around which matter could have started to clump together to form galaxies. Scientists should now see the variations in density as tiny fluctuations or anisotropies in the temperature of the cosmic microwave background. (An *anisotropy* is a difference in the physical properties of space, such as temperature and density, along one direction from the properties in another direction.)

In 1992, NASA's Cosmic Background Explorer (COBE) satellite, which just three years earlier had measured the temperature of the microwave background to an unprecedented accuracy, achieved what many astronomers consider an even greater triumph: It detected hot and cold spots in the cosmic microwave background. The COBE measurements earned the Nobel Prize in Physics for 2006 for both my NASA colleague John Mather and George Smoot of the University of California, Berkeley.

The variations are indeed minuscule — less than 10,000th of a degree K colder or hotter than the average temperature of 2.73 K. The princess who could feel a pea through a big stack of mattresses wouldn't have felt these differences. Nonetheless, these cosmic ripples are large enough to account for the growth of structure in the universe. You can sleep on that.

Mapping the universe with the cosmic microwave background

In the search to find out whether the universe is flat or saddle-shaped, scientists looked to the cosmic microwave background for answers. A flat universe would dictate that the temperature fluctuations have a particular pattern. A slew of balloon-borne and ground-based telescopes suggested that the microwave background may have this pattern.

In 2003, NASA reported that its Wilkinson Microwave Anisotropy Probe had mapped and measured the microwave background over the entire sky in sharper detail than ever. The WMAP team, led by Charles Bennett, answered most of the existing questions about the Big Bang except what made it happen and what exactly is dark energy. The team drew the following conclusions:

- ✔ The present age of the universe is 13.7 billion years. (Later the WMAP team refined this age to 13.73 billion years, and some experts put it at 13.74 billion years.)

- ✔ The cosmic microwave background originated when the universe was 379,000 years old.

- ✔ The first stars began to shine about 200 million years after the Big Bang.

- ✔ The universe is flat, consistent with the theory of inflation (see the section "Inflation: A Swell Time in the Universe," earlier in this chapter).

- ✔ The relative amounts of mass energy in the universe are as follows:

    - Normal matter (baryonic matter like that found on Earth): 4 percent

    - Dark matter (see Chapter 15): 23 percent

    - Dark energy: 73 percent

    Scientists had made rough estimates for all these quantities, but now they have precise values.

You can read all about WMAP and its findings on the Probe's official website at Goddard Space Flight Center, `http://map.gsfc.nasa.gov`. Check out the animations of the evolution of the universe and other cosmic topics on the site.

In a Galaxy Far Away: Standard Candles and the Hubble Constant

One of the longest-running questions in astronomy used to be "How old is the universe?" Now, thanks to WMAP, the Hubble Space Telescope, and other instruments, we know that the answer is 13.73 billion years.

So how did scientists figure out this magic number? They relied on information connected to the expansion of the universe: standard candles, which astronomers use to measure the distances of galaxies; and the Hubble constant, which relates galaxy distances to the rate at which the universe is expanding. I cover these topics in the following sections.

Standard candles: How do scientists measure galaxy distances?

Most strategies for measuring distance require some kind of *standard candle,* the cosmic equivalent of a light bulb of known wattage.

For instance, suppose that you believe you know the true brightness, or *luminosity,* of a particular type of star. Light from a distant source grows dimmer in proportion to the square of the distance, so the apparent brightness of a star of that same type in a distant galaxy indicates how far away the galaxy lies.

Yellowish, pulsating stars known as *Cepheid variables* remain one of the most credible standard candles for estimating the distance to relatively nearby galaxies (see Chapter 12). These youthful stars brighten and dim periodically. In 1912, Henrietta Leavitt of Harvard College Observatory detected that the rapidity with which Cepheids change their brightness is directly linked to their true luminosity. The longer the period, the greater the luminosity. A century has gone by since Leavitt's discovery, and astronomers still use Cepheids to measure distances in space.

Type Ia supernovas (see Chapter 11) are another type of standard candle. Because supernovas are much brighter than Cepheids, we can observe them in much more distant galaxies. Recent calculations of the Hubble constant employed both of these candles and got results that were in good agreement with each other and with the data from the WMAP satellite.

The Hubble constant: How fast do galaxies really move?

Cosmic age estimates have depended on a number that has held the attention of astronomers for decades: the *Hubble constant,* which represents the rate at which the universe is currently expanding. The number was named for Edwin Hubble, who found that we live in an expanding universe. In particular, he made the remarkable discovery that every distant galaxy (those beyond the Local Group of Galaxies, which I describe in Chapter 12) appears to be racing away from our home galaxy, the Milky Way.

Hubble found that the more remote the galaxy, the faster it recedes. This relationship is known as *Hubble's Law.* For example, consider two galaxies, one of which lies twice as far from the Milky Way as the other. The galaxy that resides twice as far away appears to move away twice as fast. (According to Albert Einstein's Theory of General Relativity, the galaxies themselves don't move; instead, the fabric of space in which they reside expands.)

The constant of proportionality that relates the distance of a galaxy to its recession speed is known as the Hubble constant, or H_o. In other words, the speed at which a galaxy recedes is equal to H_o multiplied by the galaxy's distance. H_o thus provides a measure of the rate of universal expansion and, by implication, its age. (If you know how far away a galaxy is now and the rate at which it has been moving, you can calculate how long it took to get that far. According to the Big Bang theory, the universe was once infinitesimal in size, and then space began to expand. The points in space where we are now and where a particular galaxy is now were once right on top of each other, but as the universe aged, the points moved apart. The time they took to move to their present distance from each other is the age of the universe.)

The Hubble constant is measured in kilometers per second per megaparsec. (One megaparsec is 3.26 million light-years.) After years of study, astronomers using the Hubble Space Telescope reported a value of 70 for the Hubble constant. That number means that a galaxy about 30 megaparsecs (about 100 million light-years) from Earth speeds away at 2,100 kilometers per second, which is about 1,300 miles per second. WMAP's findings suggest the value is 71; that's darn good agreement, although other recent observations suggest that the Hubble constant could be as large as 74.

Thanks to standard candles and the Hubble constant, astronomers now have reliable data on the current expansion rate of the universe, and we know that dark energy is increasing the expansion rate. But the nature of dark energy remains a deep, dark mystery.

The Fate of the Universe

Dark energy makes the universe expand faster and faster as time goes on. The Hubble constant thus doesn't stay constant very long; it gets bigger. In other words, the Hubble constant is more of a "Hubble inconstant."

As the universe keeps expanding at greater and greater rates, eventually other galaxies will be moving away from us at a rate greater than the speed of light. When you read the preceding sentence, you may have said, "Wait a minute! In Chapter 13, you told us that things can't go faster than the speed of light, except for tachyon particles, which may not even exist. So what's the deal with these galaxies?"

The answer is that, one day trillions of years in the future, the galaxies that will be moving away at a rate greater than light speed won't be doing that moving themselves. Remember, at the beginning of this chapter, I told you that the Big Bang "was the origin and rapid expansion of space itself." The seemingly high speeds of the galaxies that are rushing apart are not actual motions of galaxies; they're caused by the expansion of space itself. Space isn't matter, and it can go as fast as dark energy makes it expand.

When other galaxies are moving away at above the speed of light, their light will no longer reach our Milky Way galaxy. The Sun will be gone long before that — it will have used up its core hydrogen 4 billion years from now; see Chapter 11 for details) and, soon afterward, will become a red giant, shed its outer layers, and fade away as a white dwarf. But there may be other stars in the Milky Way, still going strong, with planets and maybe even intelligent beings. Those extraterrestrials won't see the galaxies whose light can't reach them. The universe, in effect, will go dark.

At one time, astronomers thought the universe would remain much as we know it far into the future. But the discovery of dark energy changed all that. As Yogi Berra famously said, "The future ain't what it used to be."

Ron Cowen, who covers astronomy and space for many publications, originally contributed this chapter. The author, Stephen P. Maran, updated it for the subsequent and current editions of Astronomy For Dummies. *All opinions expressed in this chapter are those of the author.*

Part V
The Part of Tens

The 5th Wave   By Rich Tennant

PROF. HOFFMAN SUCCESSFULLY PASSES THROUGH A WORMHOLE TO ANOTHER DIMENSION IN TIME AND SPACE.

In this part . . .

Do you ever find yourself at a social gathering desperately trying to think of something unique and interesting to say? You search your brain for some crowd-grabbing insight to make everyone in the room take notice of your remarkable powers of thought. Well, after reading The Part of Tens, you'll be ready for the next lapse in conversation. I offer you ten strange facts about space that I guarantee will garner interest. Then I fill you in on about ten major mistakes that people and the media have made — and keep making — on the topic of astronomy.

Chapter 17

Ten Strange Facts about Astronomy and Space

Here are some of my favorite facts about astronomy and, in particular, Earth and its solar system. With the following information under your belt, you may be ready to handle the astronomy questions on television quiz shows and inquiries from friends and family.

You Have Tiny Meteorites in Your Hair

Micrometeorites, tiny particles from space visible only through microscopes, are constantly raining down on Earth. Some fall on you whenever you go outdoors. But without the most advanced laboratory equipment and analysis techniques, you can't detect them. They get lost in the great mass of pollen, smog particles, household dust, and (I'm sorry to say) dandruff that resides on the top of your head. (Check out Chapter 4 for the scoop on meteorites of all sizes.)

A Comet's Tail Often Leads the Way

A comet tail isn't like a horse tail, which always trails behind as the horse gallops ahead. A comet tail always points away from the Sun. When a comet approaches the Sun, its tail, or tails, stream behind it; when the comet heads back out into the solar system, the tail leads the way. (See Chapter 4 for more information about comets.)

Earth Is Made of Rare and Unusual Matter

The great majority of all the matter in the universe is so-called *dark matter*, invisible stuff that astronomers haven't yet identified (see Chapter 15). And most ordinary or visible matter is in the form of plasma (hot, electrified gas that makes up normal stars such as the Sun) or degenerate matter (in which atoms or even the nuclei within the atoms are crushed together to unimaginable density, as found in white dwarfs and neutron stars; see Chapter 11). You don't find dark matter, degenerate matter, or much plasma on Earth. Compared to the great bulk of the universe, Earth and earthlings are the aliens. (See Chapter 5 for more about Earth's unique properties.)

High Tide Comes on Both Sides of the Earth at the Same Time

Ocean tides on the side of Earth that faces the Moon are not appreciably higher than tides on the opposite side of Earth at the same time. This may defy common sense, but not physics and mathematical analysis. (The same goes for the smaller ocean tides raised by the Sun.) See Chapter 5 for more about the Moon.

On Venus, the Rain Never Falls on the Plain

In fact, the constant rain on Venus never falls on anything. It evaporates before it hits the ground, and the rain is pure acid. (The common name for evaporating rain is *virga;* see Chapter 6 for more about Venus.)

Rocks from Mars Dot the Earth

People have found about 100 meteorites on Earth that come from the crust of Mars, blasted from that planet by the impacts of much larger objects — perhaps from the asteroid belt (see Chapter 7 for info on asteroids). Statistically, many more undiscovered Mars rocks must have fallen into the ocean or landed in out-of-the-way places where they haven't been spotted. (See Chapter 6 to find out more about Mars.)

Pluto Was Discovered from the Predictions of a False Theory

Percival Lowell predicted the existence and approximate location of the object that we now call Pluto. When Clyde Tombaugh surveyed the designated region, he discovered Pluto. But now scientists know that Lowell's theory, which inferred the existence of Pluto from its gravitational effects on the motion of Uranus, was wrong. In fact, Pluto's mass is much too small to produce the "observed" effects. Furthermore, the "gravitational effects" were just errors in measuring the motion of Uranus. (Not enough information was available about Neptune's motion to study it for clues.) The discovery of Pluto took hard work, but as it happened, it was just plain luck. And although Lowell predicted the existence of a planet, as Pluto was first termed, the International Astronomical Union has since downgraded it to dwarf planet. (See Chapter 9 to find out more about Pluto.)

Sunspots Are Not Dark

Almost everyone "knows" that sunspots are "dark" spots on the Sun. But in reality, sunspots are simply places where the hot solar gas is slightly cooler than its surroundings (see Chapter 10 for more explanation). The spots look dark compared to their hotter surroundings, but if all you can see is the sunspot, it looks bright.

A Star in Plain View May Have Exploded, but No One Knows

Eta Carinae is one of the most massive, fiercely shining stars in our galaxy, and astronomers expect it to produce a powerful supernova explosion at any time, if it hasn't already. But because light takes about 8,000 years to travel from Eta Carinae to Earth, an explosion that occurred sooner than that many years ago isn't visible to us yet. (See Chapter 11 to discover more about the life cycles of stars.)

You May Have Seen the Big Bang on an Old Television

Some of the *snow* — a pattern of interference that looks like little white spots or streaks on old black-and-white television sets — was actually radio waves the TV antenna received from the cosmic microwave background, a glow from the early universe in the aftermath of the Big Bang (see Chapter 16). When this radiation was actually discovered at the Bell Telephone Laboratories, scientists studied many possible causes of the unexpected "noise" in the radio receiver. They even investigated pigeon droppings as a possible cause but later dropped that suggestion.

Chapter 18

Ten Common Errors about Astronomy and Space

In This Chapter

▶ Perusing popular misconceptions about astronomy

▶ Correcting common news and entertainment media mistakes

*I*n daily life — reading the newspaper, watching the evening news, surfing the web, or talking to friends — you run across many misconceptions about astronomy. In this chapter, I explain the most common of these errors.

"The Light from That Star Took 1,000 Light-Years to Reach Earth"

Many people mistake the light-year for a unit of time on par with units like a day, a month, or an ordinary year. But a light-year is a unit of distance, equal to the length that light travels in a vacuum over a period of one year. (See Chapter 1.)

A Freshly Fallen Meteorite Is Still Hot

Actually, freshly fallen meteorites are cold; an icy frost (from contact with moisture in the air) sometimes forms on a frigid stone that has recently landed. When an eyewitness says that he saw a meteorite fall to the ground and that he burned his fingers on the rock, the account may be a hoax. (See Chapter 4 for more information about meteorites.)

Summer Always Comes When Earth Is Closest to the Sun

The belief that summer comes when Earth is closest to the Sun is about the most common error of them all, but common sense should tell you that the belief is false. After all, winter occurs in Australia when the United States is experiencing summer. But on any given day, Australia is the just as far from the Sun as the United States. In fact, Earth is closest to the Sun in January and farthest from the Sun in July. (See Chapter 5 for more explanation of our seasons.)

The Back of the Moon Is Dark

Some people think that the back of the Moon, which faces away from Earth (astronomers call it the "far side") is dark. Sometimes the far side *is* dark, but sometimes it is bright, and much of the time, part of it is dark and the rest of it is bright. In that way, the far side of the Moon is just like the near side, which faces Earth. When we see a full Moon, the near side is all bright and the far side is all dark. At new Moon, when the side of the Moon that faces Earth is dark, the far side of the Moon is bright — if we could see the far side then, it would look like a full Moon. For more about the Moon and its phases, check out Chapter 5.

The "Morning Star" Is a Star

The "morning star" isn't a star; it's always a planet. And sometimes two Morning Stars appear at once, such as Mercury and Venus (see Chapter 6). The same idea applies to the "evening star": You're seeing a planet, and you may see more than one. "Shooting stars" and "falling stars" are misnomers, too. These "stars" are meteors — the flashes of light caused by small meteoroids falling through Earth's atmosphere (see Chapter 4). Many of the "superstars" you see on television may be just flashes in the pan, but they at least get 15 minutes of fame.

If You Vacation in the Asteroid Belt, You'll See Asteroids All Around You

In just about any movie about space travel, you see a scene in which the intrepid pilot skillfully steers the spaceship past hundreds of asteroids that hurtle past in every direction, sometimes coming five at a time. Moviemakers just don't understand the vastness of the solar system, or they ignore it for dramatic purposes. If you stood on an asteroid smack dab in the middle of the main asteroid belt between Mars and Jupiter, you'd be lucky to see more than one or two other asteroids, if any, with the naked eye. (See Chapter 7 for more information about asteroids.)

Nuking a "Killer Asteroid" on a Collision Course for Earth Will Save Us

You come across many common errors about asteroids, and various Hollywood doomsday movies and media reports on "killer asteroids" have provided ample but unfortunate opportunities to reinforce these misunderstandings among the public.

Blowing up an asteroid on a collision course with Earth with an H-bomb would only create smaller and collectively-just-as-dangerous rocks, all still heading for our planet. A less risky idea is to attach a rocket motor to gently propel the asteroid just the slightest bit forward or backward in its orbit, steering it so that it doesn't get to the same place in space as the Earth at the same time. Even better, we might launch a so-called gravity tractor satellite to gently pull the asteroid from its original trajectory so that it misses the Earth. (I explain the gravity tractor method in Chapter 7.)

The Sun Is an Average Star

You often hear or read statements that the Sun is an average star. In fact, the vast majority of all stars are smaller, dimmer, cooler, and less massive than our Sun (see Chapter 10). Be proud of the Sun — it's like a kid from the mythical Lake Wobegon, where the children are all "above average."

The Hubble Telescope Gets Up Close and Personal

The Hubble Space Telescope doesn't snap those beautiful pictures by cruising through space until it floats alongside nebulae, star clusters, and galaxies (see Chapter 12). The telescope stays in close orbit around Earth, and it just takes great photos. It does so because it has incredibly well-made optics and orbits far above the parts of Earth's atmosphere that blur our view with telescopes on the ground.

The Big Bang Is Dead

When an astronomer reports a finding that doesn't fit the current understanding of cosmology, members of the media are prone to pronouncing, "The Big Bang is dead." (See Chapter 16 for an explanation of the Big Bang.) But astronomers are simply finding differences between the observed expansion of the universe and specific mathematical descriptions of it. The competing theories — including one that fits the newly reported data — are consistent with the Big Bang; they just differ in the details.

Part VI
Appendixes

"It's another deep space probe from Earth seeking contact with extraterrestrials. I wish they'd just include an email address."

In this part . . .

The two appendixes in this part offer information to enhance your skywatching experiences for years to come and to help you navigate through other books, articles, or online articles on astronomy and space. The first appendix offers maps to help you locate the constellations and find interesting stars. The second appendix is a glossary offering simple definitions of some astronomy terms that you can use as you enjoy and expand your skywatching hobby.

Appendix A

Star Maps

The following pages contain eight star maps — four of the Northern Hemisphere and four of the Southern Hemisphere — to start you on your starry way.

Map by Robert D. Miller

Brightness:
- 0 magnitude
- 1
- 2
- 3
- 4

Chart times (local, standard):
- 11:00 pm   August 20
- 10:00 pm   September 5
- 9:00 pm   September 21
- 8:00 pm   October 5

This map is for latitude 35 North, but it is useful throughout the continental United States.

Map by Robert D. Miller

Brightness:
✸ 0 magnitude
◆ 1
• 2
· 3
· 4

This map is for latitude 35 South, but it is useful
for most Southern Hemisphere observers.

Chart times (local, standard):
11:00 pm   August 20
10:00 pm   September 5
9:00 pm   September 21
8:00 pm   October 5

Brightness:
O magnitude
1
2
3
4

Map by Robert D. Miller

This map is for latitude 35 North, but it is useful throughout the continental United States.

Chart times (local, standard):
11:00 pm November 20
10:00 pm December 7
 9:00 pm December 21
 8:00 pm January 6

Brightness:
* 0 magnitude
* 1
* 2
* 3
* 4

Map by Robert D. Miller

This map is for latitude 35 South, but it is useful
for most Southern Hemisphere observers.

Chart times (local, standard):
11:00 pm	November 20
10:00 pm	December 7
9:00 pm	December 21
8:00 pm	January 6

NORTH

CASSIOPEIA CEPHEUS

DRACO

Little Dipper

Polaris

URSA MINOR

CORONA BOREALIS

URSA MAJOR

PERSEUS

Capella

Big Dipper

AURIGA

Pleiades

BOOTES

TAURUS

Arcturus

Castor
Pollux
GEMINI

Aldebaran

COMA BERENICES

CANCER

ERIDANUS

EAST

LEO

M44

Regulus

Procyon

Betelgeuse

ORION

WEST

VIRGO

CANIS
MINOR

M42

Rigel

Spica

Alphard

Sirius

CORVUS

HYDRA

CANIS MAJOR

LEPUS

Brightness:

✴	0 magnitude
✦	1
•	2
·	3
·	4

VELA

PUPPIS

Map by Robert D. Miller

SOUTH

Chart times (local, standard):	
11:00 pm	February 19
10:00 pm	March 6
9:00 pm	March 21
8:00 pm	April 5

This map is for latitude 35 North, but it is useful
throughout the continental United States.

Brightness:
* 0 magnitude
* 1
* 2
* 3
* 4

Map by Robert D. Miller

Chart times (local, standard):
11:00 pm	February 19
10:00 pm	March 6
9:00 pm	March 21
8:00 pm	April 5

This map is for latitude 35 South, but it is useful for most Southern Hemisphere observers.

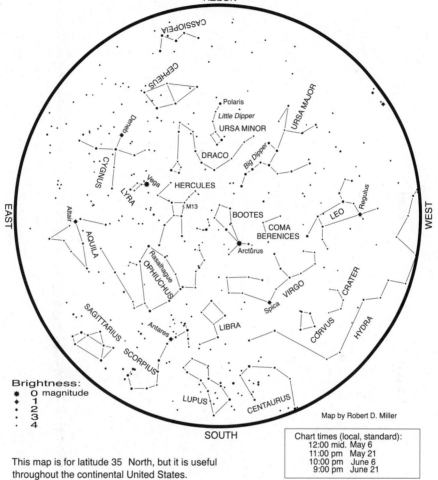

Map by Robert D. Miller

Chart times (local, standard):
12:00 mid. May 6
11:00 pm May 21
10:00 pm June 6
9:00 pm June 21

This map is for latitude 35 North, but it is useful throughout the continental United States.

Brightness:
✹ 0 magnitude
◆ 1
• 2
· 3
· 4

Map by Robert D. Miller

Chart times (local, standard):
12:00 mid. May 6
11:00 pm May 21
10:00 pm June 6
9:00 pm June 21

This map is for latitude 35 South, but it is useful
for most Southern Hemisphere observers.

Appendix B

Glossary

• •

antimatter: Matter composed of antiparticles that have the same mass but opposite electrical charge of ordinary particles.

asterism: A named pattern of stars, such as the Big Dipper, that isn't one of the 88 official constellations.

asteroid: One of many small, rocky, and/or metallic bodies orbiting the Sun.

aurora: The occurrence of a colorful display of light in the upper atmosphere of Earth or another planet, produced by the collision of electrically charged particles with gas atoms and molecules.

binary star: Two stars orbiting around a common center of mass in space; also called a binary system.

black hole: An object with a gravity so strong that nothing inside can escape — not even a ray of light.

bolide: A very bright meteor that appears to explode or that produces a loud noise.

comet: One of many small bodies made of icy and dusty matter orbiting the Sun.

constellation: Any one of 88 regions on the sky, typically named after an animal, object, or ancient deity (for example, Ursa Major, the Great Bear).

crater: A round depression on the surface of a planet, moon, or asteroid created by the impact of a falling body, a volcanic eruption, or the collapse of an area.

dark energy: An unexplained physical process that acts as if it were a repulsive force, causing the universe to expand at a greater and greater rate as time goes on.

dark matter: One or more unknown substances in space that exert gravitational force on celestial objects, which is how astronomers detect its existence.

Doppler effect: The process by which light or sound is altered in perceived frequency or wavelength by the motion of its source with respect to the observer.

double star: Two stars that appear very close to each other on the sky and that may be physically associated (a binary star) or may be unrelated to each other and at different distances from Earth.

dwarf planet: An object that orbits the Sun, isn't the moon of a planet, is massive enough that its own gravity makes it round, and hasn't cleared its orbit of other small bodies. Pluto is a dwarf planet.

eclipse: The partial disappearance (partial eclipse) or total disappearance (total eclipse) of a celestial body when another object passes in front of it or when it moves into the shadow of another object.

ecliptic: The apparent path of the Sun across the background of the constellations.

exoplanet: A planet of a star other than the Sun. Also called an *extrasolar planet.*

fireball: A very bright meteor.

galaxy: A huge system of billions of stars, sometimes with vast amounts of gas and dust.

gamma ray burst: An intense outburst of gamma rays that comes without warning from a random spot in the distant universe.

meteor: The flash of light caused by the fall of a meteoroid through Earth's atmosphere; the term *meteor* is often incorrectly used to refer to the meteoroid itself.

meteorite: A meteoroid that landed on Earth.

meteoroid: A rock in space, composed of stone and/or metal; probably a chip from an asteroid.

near-Earth object: An asteroid or comet that follows an orbit that brings it close to Earth's orbit around the Sun.

nebula: A cloud of gas and dust in space that may emit, reflect, and/or absorb light.

neutrino: A subatomic particle that has no electric charge and an extremely small mass. It can pass through a whole planet or even the Sun.

neutron star: An object only tens of miles across but greater in mass than the Sun (all pulsars are neutron stars, but not all neutron stars are pulsars).

occultation: The process by which one celestial body passes in front of another, blocking it from the view of an observer.

orbit: The path followed by a celestial body or a spacecraft.

planet: A large, round object that forms in a flattened cloud around a star and — unlike a star — doesn't generate energy by nuclear reactions.

planetary nebula: A glowing, expanding gas cloud that was expelled in the death throes of a Sun-like star.

pulsar: A fast-spinning, tiny, and immensely dense object that emits light, radio waves, and/or X-rays in one or more beams like the beam from a lighthouse.

quasar: A small, extremely bright object at the center of a distant galaxy, thought to represent the emission of much energy from the surroundings of a giant black hole.

red giant: A large, very bright star with a low surface temperature; also a late stage in the life of a Sun-like star.

redshift: An increase in the wavelength of light or sound, often due to the Doppler effect or, in the case of distant galaxies, the expansion of the universe.

rotation: The spinning of an object around an axis that passes through it.

seeing: A measure of the steadiness of the air at a place of astronomical observation (when the seeing is good, the images you view through telescopes are sharper).

SETI: The Search for Extraterrestrial Intelligence, a program of radio astronomy observations (and other observations) that seeks to detect messages from intelligent civilizations elsewhere in space.

solar activity: Changes in the appearance of (and in the radiation from) the Sun that occur from second to second, minute to minute, hour to hour, and even year to year. It includes eruptions such as solar flares and coronal mass ejections and other features such as sunspots.

spectral type: A classification applied to a star based on the appearance of its spectrum, usually related to the temperature in the region where the visible light from the star originates.

star: A large mass of hot gas held together by its own gravity and fueled by nuclear reactions.

star cluster: A group of stars held together by their mutual gravitational attraction that formed together at about the same time (types include globular clusters and open clusters).

sunspot: A relatively cool and dark region on the visible surface of the Sun, caused by magnetism.

supernova: An immense explosion that disrupts an entire star and that may form a black hole or a neutron star.

terminator: The line separating the illuminated and dark parts of a celestial body that shines by reflected light.

transit: The movement of a smaller object, such as Mercury, in front of a larger object, such as the Sun.

variable star: A star that changes perceptibly in brightness.

white dwarf: A small, dense object shining from stored heat and thus fading away; the final stage in the life of a Sun-like star.

zenith: The point on the sky that's directly above the observer.

Sky Measures

arc minutes/arc seconds: Units of measurement on the sky. A full circle around the sky consists of 360°, each divided into 60 arc minutes; each arc minute is divided into 60 arc seconds.

astronomical unit (AU): A measure of distance in space, equal to 149,597,870,700 meters or approximately 92,955,807 miles, or roughly the average distance between Earth and the Sun.

declination: On the sky, the coordinate that corresponds to latitude on Earth and that's measured in degrees north or south of the celestial equator.

light-year: The distance light travels in a vacuum (for example, through space) in one year; about 5.9 trillion miles.

magnitude: A measure of the relative brightness of stars, with smaller magnitudes corresponding to brighter stars. For example, a 1st magnitude star is 100 times brighter than a 6th magnitude star.

right ascension: On the sky, a coordinate that corresponds to longitude on Earth and that's measured eastward from the vernal equinox (a point on the sky where the celestial equator crosses the ecliptic and where the Sun is located on the first day of spring in the Northern Hemisphere).

Index

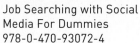

le & Mac

2 For Dummies,
Edition
1-118-17679-5

ne 4S For Dummies,
Edition
1-118-03671-6

touch For Dummies,
Edition
1-118-12960-9

OS X Lion
Dummies
1-118-02205-4

gging & Social Media

Ville For Dummies
1-118-08337-6

ebook For Dummies,
Edition
1-118-09562-1

n Blogging
Dummies
1-118-03843-7

ter For Dummies,
Edition
0-470-76879-2

dPress For Dummies,
Edition
1-118-07342-1

iness

Flow For Dummies
1-118-01850-7

sting For Dummies,
Edition
0-470-90545-6

Job Searching with Social Media For Dummies
978-0-470-93072-4

QuickBooks 2012
For Dummies
978-1-118-09120-3

Resumes For Dummies,
6th Edition
978-0-470-87361-8

Starting an Etsy Business
For Dummies
978-0-470-93067-0

Cooking & Entertaining

Cooking Basics
For Dummies, 4th Edition
978-0-470-91388-8

Wine For Dummies,
4th Edition
978-0-470-04579-4

Diet & Nutrition

Kettlebells For Dummies
978-0-470-59929-7

Nutrition For Dummies,
5th Edition
978-0-470-93231-5

Restaurant Calorie Counter
For Dummies,
2nd Edition
978-0-470-64405-8

Digital Photography

Digital SLR Cameras &
Photography For Dummies,
4th Edition
978-1-118-14489-3

Digital SLR Settings
& Shortcuts
For Dummies
978-0-470-91763-3

Photoshop Elements 10
For Dummies
978-1-118-10742-3

Gardening

Gardening Basics
For Dummies
978-0-470-03749-2

Vegetable Gardening
For Dummies,
2nd Edition
978-0-470-49870-5

Green/Sustainable

Raising Chickens
For Dummies
978-0-470-46544-8

Green Cleaning
For Dummies
978-0-470-39106-8

Health

Diabetes For Dummies,
3rd Edition
978-0-470-27086-8

Food Allergies
For Dummies
978-0-470-09584-3

Living Gluten-Free
For Dummies,
2nd Edition
978-0-470-58589-4

Hobbies

Beekeeping
For Dummies,
2nd Edition
978-0-470-43065-1

Chess For Dummies,
3rd Edition
978-1-118-01695-4

Drawing For Dummies,
2nd Edition
978-0-470-61842-4

eBay For Dummies,
7th Edition
978-1-118-09806-6

Knitting For Dummies,
2nd Edition
978-0-470-28747-7

Language & Foreign Language

English Grammar
For Dummies,
2nd Edition
978-0-470-54664-2

French For Dummies,
2nd Edition
978-1-118-00464-7

German For Dummies,
2nd Edition
978-0-470-90101-4

Spanish Essentials
For Dummies
978-0-470-63751-7

Spanish For Dummies,
2nd Edition
978-0-470-87855-2

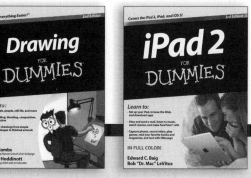

Math & Science

Algebra I For Dummies,
2nd Edition
978-0-470-55964-2

Biology For Dummies,
2nd Edition
978-0-470-59875-7

Chemistry For Dummies,
2nd Edition
978-1-1180-0730-3

Geometry For Dummies,
2nd Edition
978-0-470-08946-0

Pre-Algebra Essentials
For Dummies
978-0-470-61838-7

Microsoft Office

Excel 2010 For Dummies
978-0-470-48953-6

Office 2010 All-in-One
For Dummies
978-0-470-49748-7

Office 2011 for Mac
For Dummies
978-0-470-87869-9

Word 2010
For Dummies
978-0-470-48772-3

Music

Guitar For Dummies,
2nd Edition
978-0-7645-9904-0

Clarinet For Dummies
978-0-470-58477-4

iPod & iTunes
For Dummies,
9th Edition
978-1-118-13060-5

Pets

Cats For Dummies,
2nd Edition
978-0-7645-5275-5

Dogs All-in One
For Dummies
978-0470-52978-2

Saltwater Aquariums
For Dummies
978-0-470-06805-2

Religion & Inspiration

The Bible For Dummies
978-0-7645-5296-0

Catholicism For Dummies,
2nd Edition
978-1-118-07778-8

Spirituality For Dummies,
2nd Edition
978-0-470-19142-2

Self-Help & Relationships

Happiness For Dummies
978-0-470-28171-0

Overcoming Anxiety
For Dummies,
2nd Edition
978-0-470-57441-6

Seniors

Crosswords For Seniors
For Dummies
978-0-470-49157-7

iPad 2 For Seniors
For Dummies, 3rd Edition
978-1-118-17678-8

Laptops & Tablets
For Seniors For Dummies,
2nd Edition
978-1-118-09596-6

Smartphones & Tablets

BlackBerry For Dummies,
5th Edition
978-1-118-10035-6

Droid X2 For Dummies
978-1-118-14864-8

HTC ThunderBolt
For Dummies
978-1-118-07601-9

MOTOROLA XOOM
For Dummies
978-1-118-08835-7

Sports

Basketball For Dummies,
3rd Edition
978-1-118-07374-2

Football For Dummies,
2nd Edition
978-1-118-01261-1

Golf For Dummies,
4th Edition
978-0-470-88279-5

Test Prep

ACT For Dummies,
5th Edition
978-1-118-01259-8

ASVAB For Dummies,
3rd Edition
978-0-470-63760-9

The GRE Test For
Dummies, 7th Edition
978-0-470-00919-2

Police Officer Exam
For Dummies
978-0-470-88724-0

Series 7 Exam
For Dummies
978-0-470-09932-2

Web Development

HTML, CSS, & XHTML
For Dummies, 7th Editio
978-0-470-91659-9

Drupal For Dummies,
2nd Edition
978-1-118-08348-2

Windows 7

Windows 7
For Dummies
978-0-470-49743-2

Windows 7
For Dummies,
Book + DVD Bundle
978-0-470-52398-8

Windows 7 All-in-One
For Dummies
978-0-470-48763-1